Internet of Things in Smart Sewer and Drainage Systems

Abdul Salam

Internet of Things in Smart Sewer and Drainage Systems

Theory and Applications

Springer

Abdul Salam
Computer and Information Technology
Purdue University
West Lafayette, IN, USA

ISBN 978-3-031-48108-6 ISBN 978-3-031-48109-3 (eBook)
https://doi.org/10.1007/978-3-031-48109-3

This Springer imprint is published by the registered company Springer Nature Switzerland AG
The registered company address is: Gewerbestrasse 11, 6330 Cham, Switzerland

Paper in this product is recyclable.

This book is dedicated to Umbreen, Alizay, Anzalna, Naba, *and* Mohammad.

Preface

Internet of Things in Sewer and Drainage Systems adds new empirical and analytical results, technology, and innovations to the body of knowledge. It enables researchers and industry professionals from the public and private sectors across the globe to accelerate research on sewer systems management using IoT-based sensing, and networking technologies. It presents diverse and unique next-generation IoT applications in sewer and storm drainage areas. It is an excellent book for graduate students, academic researchers, and industry professionals, involved in combined sewer overflow. It is also useful for professionals in establishing their technical competence.

This multidisciplinary book provides insights into the applications of technological innovations of Internet of Things to the combined sewer overflows (CSO) and storm water management (SWM) systems. It explores technical challenges and presents recent results to improve sewer and drainage system management using wireless underground communications and sensing in Internet of Things. It addresses both existing sensing network technologies and those currently in development in three major areas of CSO: combined sewer overflows management, subsurface sensing, and antennas in the layered medium. It explores new applications of Internet of Things in sewer systems to improve public health, foster economic growth, enhance environmental quality, and responsibility for the community.

Internet of Things in Sewer and Drainage Systems is an essential reference book for advanced students on courses in combined sewer overflows (CSO) and storm water management (SWM) systems. It will also be of interest to researchers, communication engineers, system and network planners, technical mangers, and other professionals in these fields. There is no book in the market which explores relationships between wireless communications and sensor technologies across many different areas of sewer and drainage systems by using Internet of Underground Things. Currently, this audience gets information about this topic from different sources (e.g., IEEE conferences, COMSOC tutorials, IEEE journals and transactions articles, and web forums). The purpose of this book is to transform information from these scattered sources into a comprehensive and easily accessible knowledge body.

Internet of Things in Sewer and Drainage Systems is split into four parts: Novel Sewer Overflow Sensing Techniques; Underground Antenna and Radio Interface Technologies in Sewer Systems; Internet of Things Advancements in Sewer Systems; Wireless Applications in Combined Sewer Overflows (CSO) and Storm Water Management (SWM) Systems; and Physical Layer for Wireless Communications. It starts by introducing emerging technologies in wastewater systems including wireless communications in layered medium before moving on to cover propagation models in pipelines, soil properties, and beyond. Chapters introduce the work in sensor systems, antenna arrays, and MIMO; capacity and path loss analysis; and empirical and statistical channel models in smart wireless technology for the effective management of combined sewer overflows. It investigates how underground sensors and antenna design; cross layer and environment-aware protocol design; and energy harvesting and power transfer are enabling innovations in sewer and drainage systems. This valuable resource:

- Provides a comprehensive reference for all aspects of sewer and drainage systems
- Focuses on fundamental applications of digital technologies of sewer system management in an easy language that is understandable by a wide audience
- Includes applications of Internet of Things technologies to assess condition of the sewer mains, manholes, and surface facilities
- Features on research developments and open research challenges in subsurface sensing in combined sewer overflows (CSO)
- Includes advanced treatment of IoT custom applications in preventive maintenance of sanitary sewers
- Provides a detailed set of path loss, antenna, and wireless underground channel measurements in novel storm drains and sanitary sewers testbeds
- Provides MATLAB programs for models and analysis for download
- Finally, draws these studies together and points to future developments.

The graduate students, wastewater technicians, academic researchers, sanitary engineers, industry professionals and mangers, superintendents, and environmental specialist will need and buy Internet of Things in Sewer and Drainage Systems. There are many such individuals in academia and sewer system management business, which is mostly targeted to the applications of technology in big and small sanitary systems. Moreover, the use of digital technology in diverse topographic sewer and drainage fields is also expanding. To effectively utilize enormous data being generated from the wastewater overflow monitoring systems, this book has introduced technologies which are useful for data collection, design of decision tools, interpretation, and real-time decision making in the field. The IEEE Societies which are potential buyers of the books are IEEE Communications Society, IEEE Antennas and Propagation Society, IEEE Computer Society, IEEE Electromagnetic Compatibility Society, IEEE Information Theory Society, IEEE Instrumentation and Measurement Society, IEEE Signal Processing Society, IEEE Sensors Council, IEEE Systems Council, IEEE Civil Engineering Society, IEEE Geoscience and

Remote Sensing Society, and IEEE Microwave Theory and Techniques Society. The potential non-IEEE societies interested in *Internet of Things in Sewer and Drainage Systems* are American Society of Sanitary Engineering, American Society for Engineering Education, and public works departments across the globe. The key meetings and conferences that would most welcome this are book IEEE INFOCOM, SECON, ICC, ICNSC, Globecom, Water for Food Global Institute Forum, National Soil Moisture Network, International Conference on Sanitary Systems, and Science Societies Annual Conference. The wastewater facility management coordinators, wastewater collection equipment manufacturers, embedded systems industry, radio and sensor manufactures, environmental specialists, sanitary engineers, and city planners are most related to this books subject. The research on the subject is generally published in *IEEE Transactions of Antennas and Propagation*, *Civil Engineering Journal*, *Ad Hoc Networks Journal*, *IEEE Transactions on Geoscience and Remote Sensing*, *IEEE Communication Magazine*, *IEEE Surveys and Tutorials*, *Journal of the Sanitary Engineering*, and *IEEE Transactions on Wireless Communications*. It is also possible to adopt this book classroom graduate textbook. It can be used in course titles such as Internet of Things, Wireless Communications, Cyber-Physical Systems, Embedded Systems, Smart Cities, Sewer Engineering, and Sustainably.

West Lafayette, IN, USA — Abdul Salam
December 2023

Acknowledgments

I could never be grateful enough to my wife *Umbreen Abdul Salam* and my children *Alizay Meerub Bilal*, *Anzalna Akhtar Sajid*, *Naba Salam*, and *Mohammad Yahya Hassan* for their unqualified encouragement that has meant so much to me.

Contents

About the Author

Abdul Salam is an Assistant Professor with the Department of Computer and Information Technology, Purdue University. He holds a Ph.D. from University of NebraskaLincoln. He is the Director of the Environmental Networking Technology Laboratory at Purdue University. He has also taught at the Bahauddin Zakariya University, Multan and Islamia University, Bahawalpur. He served in Pakistan Army for 9 years in a number of command, staff, and field roles. He held the principal position at the Army Public School and College, Thal Cantonment. He has won several awards including the ICCCN 2016 Best Student Paper Award, the Robert B. Daugherty Water for Food Institute Fellowship, the Gold Medal MS (CS) on securing first position in order of merit, and the 2016–2017 Outstanding Graduate Student Research Award from the University of Nebraska–Lincoln. He is the author of three books, *Internet of Things for Sustainable Community Development*, *Signals in the Soil*, and *Internet of Things in Smart Sewer and Drainage Systems*. He has published over 60 research articles in major journals and international conferences. His recent research focuses on underground soil sensing, wireless communications, Internet of Underground Things in digital agriculture, sensor-guided irrigation systems, and vehicular communications. He has served as an Associate Editor for the *IEEE GRSS Remote Sensing Code Library* from 2016 to 2018. He serves as the Associate Editor of the *Advanced Electromagnetics Journal* and *Array* (Elsevier). Dr. Salam received his B.Sc. and M.S. degrees in Computer Science from Bahauddin Zakariya University, Multan, Pakistan, in 2001 and 2004, respectively, the M.S. degree in Computer Engineering from UET, Taxila, Pakistan, in 2012, and the Ph.D. degree in Computer Engineering from the Cyber-Physical Networking Laboratory, Department of Computer Science and Engineering, University of Nebraska–Lincoln, Lincoln, NE, USA.

Chapter 1
IoT in Stormwater Management

Check for updates

1.1 Introduction

Smart water systems make it easier to deploy cost-effective sensors and wireless communication setup in the watersheds. This coupled with Internet of Things (IoT) outputs a scalable system with an ability to manage thousands of devices [20]. With a glut of sensors in watersheds, researchers can have a system-level comprehension of the system allowing them to make decisions and alert public about any unforeseeable event in advance. Moreover, such system also allows for regulation of water level to minimize flooding [2] and distributed control of the watershed. Distributed control allows operators to control the stormwater [19] tank outflows over different sites (10s to 100s) in city. This allows to reduce the chances of water flooding and usage of existing ponds to the full capacity (Fig. 1.1).

To summarize, development of the above-mentioned smart water system is not limited by technology anymore. However, the development is hampered by: (1) too much dependence on proprietary technologies, (2) lack of working and proven case studies, and (3) the absence of solutions that are specifically designed and tested for water application. Therefore, there is a need of solutions that, in addition to technology, also utilizes domain knowledge of water science. Next section reviews such existing systems. After that, open-storm framework is presented as end-to-end system that aims to enable a reliable and scalable watershed control system.

1.2 Related Work

1.2.1 SCADA

Supervisory control and data acquisition (SCADA) systems consist of devices, protocols, and software for remote control and management of water resources [11].

A. Salam, *Internet of Things in Smart Sewer and Drainage Systems*,
https://doi.org/10.1007/978-3-031-48109-3_1

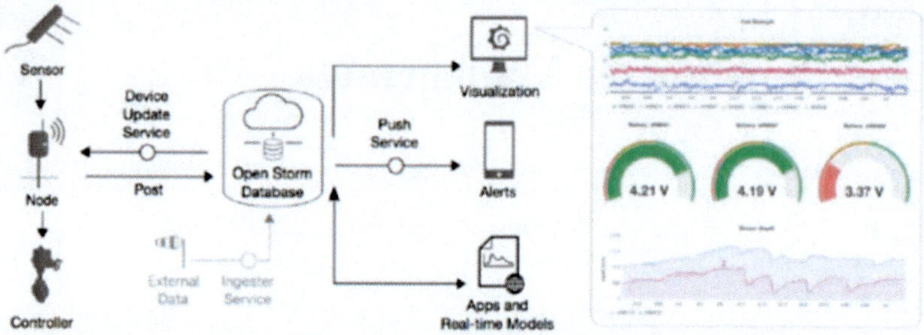

Fig. 1.1 The open-storm stack. The hardware layer (left) comprises the sensor node along with auxiliary sensors and actuators. The cloud services layer (center) includes the database backend, along with a series of publication and subscription services for controlling sensor node behavior and interfacing with applications. The applications layer (right) allows for real-time supervision and control of field-deployed devices. The rightmost panel shows an example dashboard, including sensor feeds and network status visualizations

SCADA systems use the service quality parameters (flows, pressure, etc.) to regulate pumps and valves both, manually or automatic, in real time. SCADA legacy systems have extensively deployed in water monitoring utilities; however, they are limited by interoperability, security, and scalability.

Interoperability is the major limitation of legacy SCADA systems meaning that it is possible that systems in different municipalities are incapable of intercommunication [7, 13]. This makes it hard to control and coordinate watershed activities in different areas. Another issue with SCADA system is scalability and power requirement. Traditionally, SCADA systems are limited to only centralized implementation with sensors installed on selected locations [11]. For decentralized implementation (natural river systems and stormwater networks), power requirements of system are cost prohibitive. Finally, most of the tradition SCADA systems, deployed decades ago, are not secured [9]. These legacy systems do not have any authentication protocol and lack in-built authentication capabilities. Therefore, SCADA systems have to be separated from public Internet Networks; however, remote attacks are still possible.

SCADA systems are the most popular in water management utilities; however, they need significant improvement in expanding, security, and integration with modern and old systems.

1.2.2 *Wireless Sensor Networks*

Advancements in Wireless Sensor Networks (WSNs) have complete revolutionized cost and power consumption of monitoring systems [3]. Flood monitoring system is an active application of WSNs in water science. There are many riverine flood monitoring systems with a varying number of sensor nodes implemented by various researchers. For example, 15-node system is implemented in the UK [8, 18], a 4-node system in Honduras [1], a 3-node system in Massachusetts, and a largest 200-node system in Iowa Flood Information System (IFIS) [4]. In addition to these large systems, some systems [10, 14] also perform flash flood monitoring.

Some of the WSNs also project working on achieving real-time control. For example, in [5, 15], Web-enabled sensor nodes are being used to predict weather forecast for rain harvesting in selected locations. In South Bend, Indiana, a large-scale (120-nodes) WSN is being used to modulate independent water flows into combined water sewerage system of the city [12].

Many researchers have realized the potential of WSN in real-time monitoring applications; however, they have yet implemented the projects on small scale. Moreover, they rely only on previous research project, and there is no complete documentation for end-to-end platform [1]. As a result, each researcher starts their WSN project from scratch that is very time-consuming. Therefore, there is a need to create a *how-to* guide on developing and implementing open-source WSNs systems.

1.3 The Open-Storm PLATFORM

The open-storm platform is the only end-to-end open-source platform equipped with various sensing, control, and cloud services for water resource management. It also comes with the extensive documentation and guide to sensor network deployment. The documentation is available at open-storm.org. This reference guide makes it easier for the new researchers to deploy their own sensors and control.

Open-storm framework is comprised of 3 layers: (1) Hardware, (2) Cloud Services, and (3) Applications. Hardware consists of all the hardware devises, e.g., sensors and microprocessors, etc. Cloud Services include processing utilities to interact with the devices through applications. Application layer includes algorithm to interact with hardware devices. All these layers are shown in Fig. 4.1.

1.3.1 Hardware

Hardware layer consists of: (a) sensor nodes and (b) sensors and actuators. Sensor node is an embedded computer with wireless capabilities. The function of this sensor node is to communicate information from sensors and communicate with servers. The sensor node is designed to minimize power consumption by operating in sleep mode for most of the time. Open-storm nodes use cellular 2G, 3G, and 4G technologies to wirelessly transfer data to and from servers. There is also a power regulation system attached with the sensor node along with solar panels and charger. Solar panels allow to battery charging for a long period of time so that sensor node keeps operating without any routine maintenance. Further details about the sensor node can be found at *open-storm.org/node*.

In addition to sensor nodes, open storm consists of a variety of sensors such as soil moisture sensors, water-level sensors, and water quality sensors, etc. Moreover, it has ability to integrate new sensors by adding drivers to firmware of sensor node. There are some internal sensors that are used to monitor the health parameters, such as battery voltage and reception strength, etc., of devices. These measurement are helpful in network troubleshooting and diagnostics. Moreover, it also provides functionality to control sensors remotely through web services. In addition to sensors, open storm also comes with the actuators to physically control the devices in the field. For example, Butterfly valve is used to control the discharge from storage elements. Further details about the sensors and actuators supported by the system can be found at *open-storm.org/node*.

1.3.2 Cloud Services

Sensors can operate locally; however, in order to achieve system-level control and supervision, it is important to have integration with cloud services. Cloud service layer is responsible for wireless data transmission, data storage, visualization, and remote management of deployed devices, and integration with control algorithms and real-time models. These services can be hosted locally, on remote server and on popular cloud services such as Microsoft Azure or Amazon Elastic Compute Cloud (EC2).

The cloud service layer works from two perspectives: (1) Deployed Devices and (2) Server. From devices' perspective, sensor node takes all the measurements from sensors and sends it to the server's database. It also gets the instructions from the server's database to perform desired actions based on readings. From server's perspective, application gets the readings from sensor and inputs those readings to user-defined applications to control the hardware remotely.

Open storm uses a time series database, InfluxDB, as backend. It is secured using basic username/password authentication and Transport Layer Security encryption (TLS/SSL) protocol. Database performs both, storage and communication, functions. The sensor measurement data are sent through secure web connection using HTTPS protocol. Database also facilitates secure bi-directional communication between node and cloud application. Server sends commands for node to database, and sensor nodes fetch that data from database to execute required decision. To summarize, system has a safe and secure, easily maintainable, and scalable database backend.

1.3.3 Application

Application layer is responsible for processing and analyzing the data. It also includes user interface and remote control of sensor nodes. Applications subscribe to database for working. The subscription includes the permission to read, write, and trigger an event to the database as per the user-specified condition. This allows for development of different types of applications. For example, in Fig. 4.1, open storm provides web interface for visualization purposes. It can be seen in Fig. 4.1 that dashboard has three parameters for visualization: strength of cellular connection is at the top, time–depth readings from sensor node in the bottom, and battery voltage monitoring at the center. Moreover, it also gives remote access by allowing users to change the parameters of sensor nodes through a web interface [6].

Open storm also provides automation to perform operations on user-specified conditions via push alerts, either confidential or broadcast, through emails or text messages. Push alerts can also be used to notify any hazardous conditions, network health, below threshold voltage levels and below threshold throughput. Moreover, alert also helps to automate sampling in response to weather forecasts. Further details about the sensors and actuators supported by the system can be found at *open-storm.org/cloud*.

1.4 Case Study: Flood Monitoring

DallasFort Worth (DFW) metroplex has been reported as one of the most flood-prone areas in the United States and results in 17 deaths per year in Texas [16]. There has been many solutions [3, 8, 11] proposed to solve the problem, but all of them lack timely prediction and communication of the flood risk. Therefore, open storm is used to solve this problem. Alert system of the open-storm framework aims to improve disaster response.

To this date, predicting flood estimates has been the most challenging problem due to the absence of spatial and temporal data. Different studies recommend different spatial and temporal resolutions for estimation. For example, some researcher recommends using a resolution of 500 meters and some recommends of temporal resolution [17]. However, currently, DFW metroplex has a capability of 1 rain gage per 600 km^2 that is too sparse for accurate prediction of flood flash. A wide-area network of flood monitoring system is built. The main idea is to use a large inexpensive array of depth sensors to monitor runoff at level as minuscule as roadways, culverts, and creeks. The network is scaled to such a large scale, using inexpensive equipment, that it was not possible to integrate with state-of-the-art stage monitoring system. A total of 40 sensor nodes were allocated to DFW flood monitoring with a total cost of 20,000 USD.

References

1. E.A. Basha, S. Ravela, D. Rus, Model-based monitoring for early warning flood detection, in *Proceedings of the 6th ACM Conference on Embedded Network Sensor Systems*, pp. 295–308 (2008)

2. A. Bronstert, D. Niehoff, G. Bürger, Effects of climate and land-use change on storm runoff generation: present knowledge and modelling capabilities. Hydrological Processes **16**(2), 509–529 (2002)
3. A. Cerpa, J. Elson, D. Estrin, L. Girod, M. Hamilton, J. Zhao, Habitat monitoring: Application driver for wireless communications technology. ACM SIGCOMM Comput. Commun. Rev. **31**(2 supplement), 20–41 (2001)
4. I. Demir, W.F. Krajewski, Towards an integrated flood information system: centralized data access, analysis, and visualization. Environ. Modell. Softw. **50**, 77–84 (2013)
5. EPA, Clean watersheds needs survey 2012 report to congress (2012)
6. EPA, The importance of operation and maintenance for the long-term success of green infrastructure: A review of green infrastructure O&M practices in ARRA clean water state revolving fund projects us (2015)
7. R.J. Hawley, G.J. Vietz, Addressing the urban stream disturbance regime. Freshwater Sci. **35**(1), 278–292 (2016)
8. D. Hughes, P. Greenwood, G. Blair, G. Coulson, P. Grace, F. Pappenberger, P. Smith, K. Beven, An experiment with reflective middleware to support grid-based flood monitoring. Concurren. Comput. Pract. Exp. **20**(11), 1303–1316 (2008)
9. V.M. Igure, S.A. Laughter, R.D. Williams, Security issues in SCADA networks. Comput. Secur. **25**(7), 498–506 (2006)
10. R. Marin-Perez, J. García-Pintado, A.S. Gomez, A real-time measurement system for long-life flood monitoring and warning applications. Sensors **12**(4), 4213–4236 (2012)
11. L.W. Mays, *Water Distribution System Handbook* (McGraw-Hill Education, New York, 2000)
12. L. Montestruque, M.D. Lemmon, Globally coordinated distributed storm water management system, in *Proceedings of the 1st ACM International Workshop on Cyber-Physical Systems for Smart Water Networks*, pp. 1–6 (2015)
13. R. Powell, K.S. Hindi, *Computing and Control for the Water Industry* (Research Studies Press, Taunton, 1999)
14. C.H. See, K.V. Horoshenkov, R.A. Abd-Alhameed, Y.F. Hu, S.J. Tait, A low power wireless sensor network for gully pot monitoring in urban catchments. IEEE Sensors J. **12**(5), 1545–1553 (2011)
15. B. Sercu, L.C. Van De Werfhorst, J.L.S. Murray, P.A. Holden, Sewage exfiltration as a source of storm drain contamination during dry weather in urban watersheds. Environ. Sci. Technol. **45**(17), 7151–7157 (2011)
16. H.O. Sharif, T.L. Jackson, Md. Moazzem Hossain, D. Zane, Analysis of flood fatalities in Texas. Nat. Hazards Rev. **16**(1), 04014016 (2015)
17. J.A. Smith, M.L. Baeck, K.L. Meierdiercks, A.J. Miller, W.F. Krajewski, Radar rainfall estimation for flash flood forecasting in small urban watersheds. Adv. Water Resour. **30**(10), 2087–2097 (2007)
18. P. Smith, K. Beven, W. Tych, D. Hughes, G. Coulson, G. Blair, The provision of site specific flood warnings using wireless sensor networks (2009)
19. C.J. Vörösmarty, P.B. McIntyre, M.O. Gessner, D. Dudgeon, A. Prusevich, P. Green, S. Glidden, S.E. Bunn, C.A. Sullivan, C.R. Liermann, et al., Global threats to human water security and river biodiversity. Nature **467**(7315), 555–561 (2010)
20. B.P. Wong, B. Kerkez, Real-time environmental sensor data: An application to water quality using web services. Environ. Modell. Softw. **84**, 505–517 (2016)

Chapter 2
Storm and Stream Technologies

2.1 Introduction

Real-time control is an inexpensive way of increasing performance efficiency of stormwater systems and replacing traditional steel-and-concrete approach of building infrastructure. The major idea behind such technology is wireless communication modules, micro-controllers, and control theory [1]. Various researchers have already investigated different solutions for real-time monitoring of stormwater control. In [6], authors have modified existing retention basins by adding real-time controllers and decreasing pollutant. An adaptive control strategy is proposed in [9–13] for the rainwater harvesting in New York with an efficiency rate of **35**–60% more than conventional system. A coordinated stormwater system is underrated and benefits are not delineated. It can be understood from the example of coordinated sewer controls in [7] and [8]. Both these studies optimized sewer control operations using sensors, actuators, and rule-based control logic. The lesson can be learned from these studies, and similar benefits can also be replicated from a coordinated stormwater management [8].

This chapter proposes an efficient stormwater discharge using a network of sensors and valve connected through Internet. It is shown that desirable flow regimes can be achieved by regulating two parallel retention basins [3]. The study is comprised of four phases: *Phase I* builds an existing real-time stormwater control system from [1] with new methods of adding retention basins and controlling them using scheduling application. *Phase II* involves characterizing the control network to shape downstream hydrograph using two differently sized impulse waves in retention basins. The characterization is done by measuring magnitude, decay, and travel time of the waves. *Phase III* uses data from *Phase II* to determine the input needed for generating flat hydrograph. *Phase IV* shows how timing of control input can be tweaked to produce synchronized or de-synchronized pulses at target location. On contrary to traditional systems, this control network uses open-source hardware and software [14–23]. The purpose of using open-source stack is to make

A. Salam, *Internet of Things in Smart Sewer and Drainage Systems*,
https://doi.org/10.1007/978-3-031-48109-3_2

it available for implementation for everyone including municipalities. Therefore, an operator manual's is prepared with a combination of documentation from open-storm.org.

2.2 Study Area and Technologies

2.2.1 *Study Area*

This study is investigated in Mallets Creek Watershed, Ann Arbor, Michigan, having an area of 26.7 km^2 with streams of combined length more than 16 km ultimately draining into Huron River and finally into Great Lakes. There are no lakes in creekshed but man-made basins to stabilize the flows in creekshed.

In order to test the effect of real-time control, a control network is established over four sites of the creek (Fig. 2.1). Water flows from retention basin at site A down 1.4 km away to wetland at cite C. From cite C, after 3 km, water merges with flows from smaller retention basins at site B. Finally, combined water streams exit the creek from site D. Control valves are installed at site A and site B for regulation at creek.

2.2.2 *Technologies and Architecture*

A wireless sensing and control network is deployed to measure the flows with creekshed. Open-storm documentation from [1] is used to build this network. The hardware layer uses ultra-low-power micro-controller (Cypress PSoC), CDMA

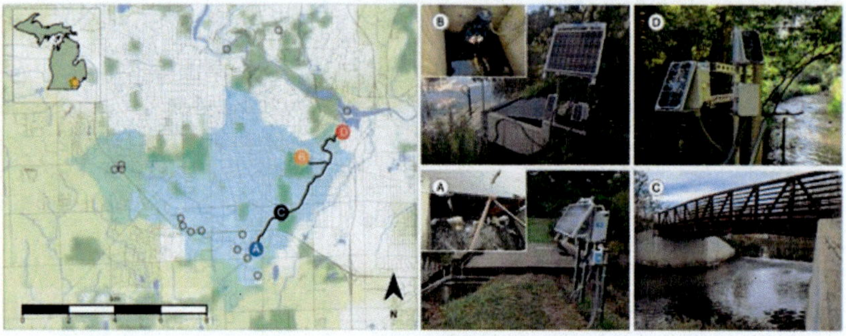

Fig. 2.1 Overview of the study area. The map (left) shows the location of relevant control and sensor sites, additional sensor sites (light gray), flow paths between each site (dark gray), and the contributing area of the watershed (light blue). Site images (right) show the two control sites (A & B) along with two downstream sensor locations (C & D)

cellular modem (Telit DE910) for Internet connectivity, and 3.7V lithium-ion battery to meet the power need of the network. Sensor node cost 500 USD and comes with ultrasonic depth sensors (MaxbotixMB7384). Two control valves, each having a diameter of 0.3 m, are retrofitted at basins at site A and site B. These valves are controlled by controllers and powered by 12 V rechargeable batteries. Total cost of each site is estimated to be 3500 USD [24–27].

Polling scheme is used to implement the control of valves and sensors. To make the setup power efficient, nodes take turns and go to long sleep cycles. At each wake-up cycle, nodes transfer readings, secured by authentication, to database using web interface (HTTP request). In case something goes bad, the operations can be re-scheduled and canceled. To add flexibility to the system, database system can also be integrated with application programming interfaces (APIs). This allows for the integration with existing application such as weather forecast [1].

A simple Python web application is written to control creekshed. There are two modes of the application: (a) manual and (b) automatic. In automatic mode, it fetches data from public measurement stations; commands are then written in response to the data gathered from the sensors. Manual mode has a predefined set of commands that are used and executed by the devices deployed in the field. Manual mode is used for this study [28–41].

2.3 Characterizing Control Actions

This section focuses on studying the capability of each site in shaping downstream flows in terms of travel time *P* and decay time *D*. The process involves having a flux of stream released from each stormwater basin at different durations (1-h release, 4-h release, and 48-h release) and then measuring the downstream waves resulted from these pulses of stream (Fig. 2.2). It was observed that 48-hr release emptied the basin at site A, and it takes 3.5 h for a wave to start rising to peak and approximately 6–8 h to reach peak. The decay times for 1-h release, 4-h release, and 48-h release were 6 h, 18 h, and 44 h, respectively. Figure 2.3 shows the results from this experiment. The change in flow from site A was recorded to be 0.14 m^3/s.

Figure 2.4 plots the results from experiments for characterization of hydraulic head of basins. Three 1-hr streams were released from B without refilling the basins between the streams. It was observed that hydraulic head of the basin decreased with each released pulse resulting smaller waves at the other end in response to each release interval.

It was observed that basin B, even though smaller in size, showed comparable results of change in flow as of basin A. It might be possible because: (a) B is closer to the output outlet resulting in less hydraulic pressure for the wave, and (b) site B is elevated causing flows to exit faster than site A. To make things simpler, it was decided to ignore the nonlinearities for the control purpose that would, otherwise, make things more complex by adding various other parameters such as variations in basin head and travel times, etc., to consider.

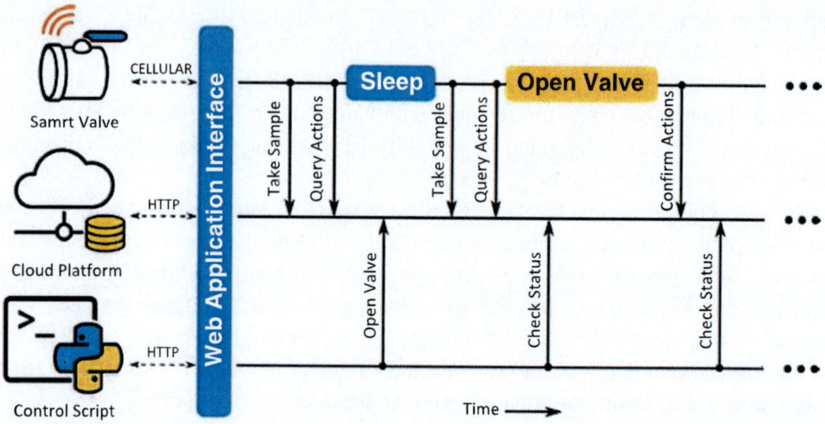

Fig. 2.2 Control system architecture. Field-deployed nodes use a polling system to download and execute commands issued from a remote server. Control actions can be specified manually, or through automated web applications and scripts

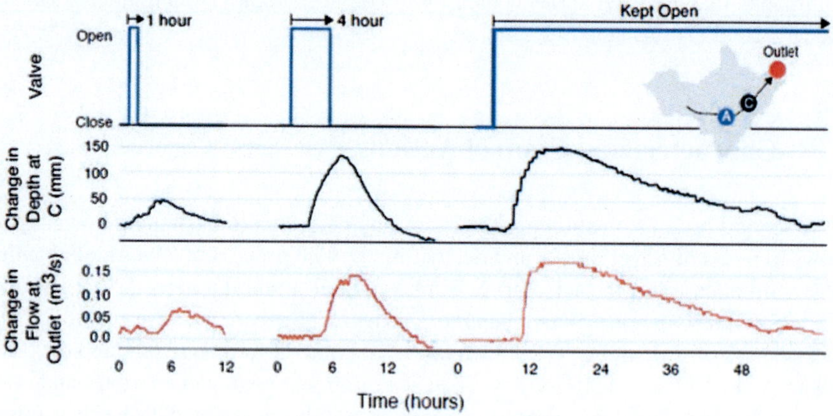

Fig. 2.3 Characterization of control actions from site A. In the first two experiments, the valve at site A is opened for 1-h and 4-h durations. For the third experiment, the valve is held open indefinitely. The resulting waves travel through a constructed wetland (site C) before arriving at the outlet of the watershed. Wave depth (black line) is measured at the wetland, while flow rate (red line) is measured at the outlet

These experiments are the first step toward implementing complex controls at outlet of the watershed. It was found that linear system approximation was enough to generate control strategies and desired waveform with a varying combination of input signals. Next, it is discussed how valve release can be timed to generate waves at the outlet [42–55, 64].

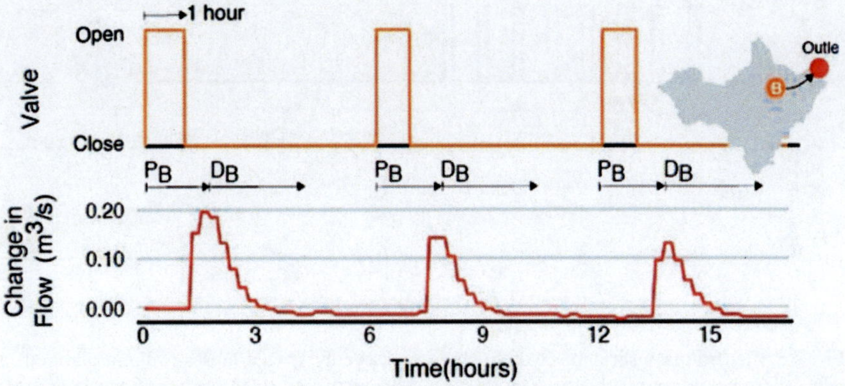

Fig. 2.4 Characterization of control actions originating from site B. Three subsequent pulses are released. While the duration of each control pulse is the same (1 h), the magnitude of the flow at the outlet decreases because the hydraulic head (pressure) in the basin is reduced with each release

2.4 Set-Point Hydrographs

A stormwater hydrograph usually has a rapid rise, followed by a peak that is further followed by a decay. This phenomenon is also seen in retention basins in Figs. 2.3 and 2.4). Peak flow rate exceeding the downstream capacity results in floods. Moreover, urban streams cannot exceed the threshold flow rate; otherwise, it becomes unstable and may result in ecological damage, if exceeded regularly [2]. This may also cause movement of sediments containing pollutants, hence resulting in contamination of water [5]. Therefore, a specific active control method is needed to stabilize the flow within urban watershed [56–59].

To that end, this study attempts to flatten the hydrograph by keeping the flow values constant and not let it exceed a threshold value. The set-points are chosen randomly, and *pulse-width modulation* concept is used to keep the flow rate constant in this experiment. The experiment involves releasing of multiple individual pulses in such a way that duration of rising peak of new pulse overlaps with the receding peak of previous pulse. Due to dispersion of the pulses down the control network, these pulses will superimpose on each other resulting in a near-to constant flow.

Although the hydrograph is not completely flat (Fig. 2.5), however, it can be seen that output remains almost flat not exceeding the threshold rate of 0.04 m^3/s in response to the input of multiple 30-minutes wave pulses. This shows that active modulation of valve can be used to achieve constant flow rate that would not be possible in case of passive infrastructure. A real-world application of this method can be to drain watershed without exceeding threshold flow rate.

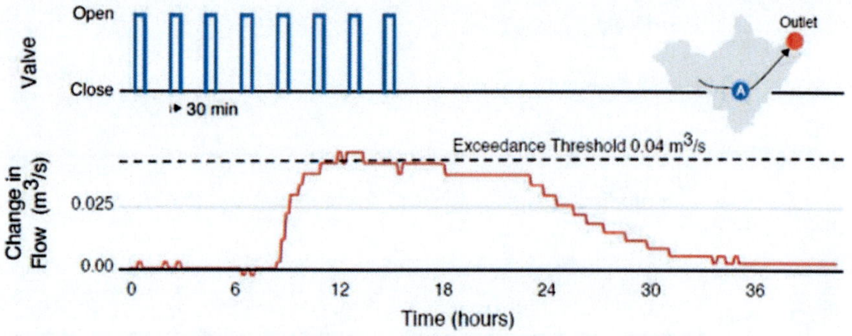

Fig. 2.5 Generating a set-point hydrograph. Small, evenly spaced pulses (30-min duration) are released from the controlled basin. The pulses disperse as they travel through the 6-km-long stream, leading to a relatively flat response at the outlet of the watershed

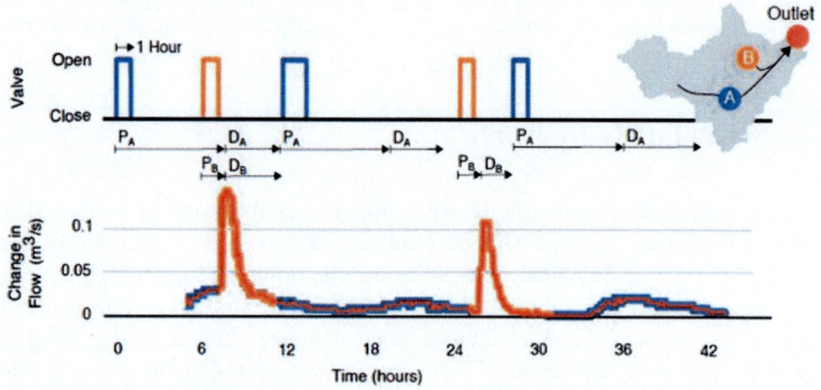

Fig. 2.6 Superposition and interleaving of waves from retention basins A and B. Overlapping waves (coincident peaks) are generated from hours 6 to 12. Interleaved waves (off-phase peaks) are generated from hours 18 to 44

2.5 Coordinated Releases Between Multiple Control Sites

In this section, release from multiple sites is coordinated to see the precision level that can be achieved for a constant flow rate. There are two retention basins A and B, and both, synchronized and interleaved releases, are investigated. There were two phases of experiments: first phase included overlaps from hydrograph peaks from both sites, and second phase included flow stream from one site arrives when flow from preceding site recedes [60, 61].

Figure 2.6 shows the results from this experiment. Overlapping waves occur between 6–15 hrs and interleaved between 15–44 hrs. For overlapping, overall flow change is nearly equal to the sum of change in component waves, i.e., exhibiting linear relation for superpositioned waves. As discussed in previous section, interleaved pattern can be used to condition the flow rates to not exceed a threshold value. Moreover, flows coming from different sources burden downstream by combining into large flows contributing to erosion. Therefore, interleaving flows can ease up the capacity of the system before the next flow, hence avoiding adverse conditions [62, 63].

This experiment allows a very precise control over the stormwater basins and helps out in coordinating releases from multiple to achieve targeted flow rates. This all is done using cost-efficient control and sensor network. This experiment sets the stage for future work to devise complex control strategies that can be used to achieve more refined control of watershed.

References

1. M. Bartos, B. Wong, B. Kerkez, Open storm: a complete framework for sensing and control of urban watersheds. Environ. Sci. Water Res. Technol. **4**(3), 346–358 (2018)
2. B.P. Bledsoe, Stream erosion potential and stormwater management strategies. J. Water Resour. Plann. Manag. **128**(6), 451–455 (2002)
3. B. Kerkez, C. Gruden, M. Lewis, L. Montestruque, M. Quigley, B. Wong, A. Bedig, R. Kertesz, T. Braun, O. Cadwalader, et al., Smarter stormwater systems (2016)
4. A. Konda, A. Rau, M.A. Stoller, J.M. Taylor, A. Salam, G.A. Pribil, C. Argyropoulos, S.A. Morin, Soft microreactors for the deposition of conductive metallic traces on planar, embossed, and curved surfaces. Adv. Funct. Mater. **28**(40), 1803020 (2018)
5. A.M. Michalak, E.J. Anderson, D. Beletsky, S. Boland, N.S. Bosch, T.B. Bridgeman, J.D. Chaffin, K. Cho, R. Confesor, I. Daloğlu, et al., Record-setting algal bloom in Lake Erie caused by agricultural and meteorological trends consistent with expected future conditions. Proc. Natl. Acad. Sci. **110**(16), 6448–6452 (2013)
6. J.R. Middleton, M.E. Barrett, Water quality performance of a batch-type stormwater detention basin. Water Environ. Res. **80**(2), 172–178 (2008)
7. A.L. Mollerup, P.S. Mikkelsen, D. Thornberg, G. Sin, Controlling sewer systems–a critical review based on systems in three EU cities. Urban Water J. **14**(4), 435–442 (2017)
8. L. Montestruque, M.D. Lemmon, Globally coordinated distributed storm water management system, in *Proceedings of the 1st ACM International Workshop on Cyber-Physical Systems for Smart Water Networks*, pp. 1–6 (2015)
9. U. Raza, A. Salam, On-site and external power transfer and energy harvesting in underground wireless. Electronics **9**(4), 681 (2020)
10. U. Raza, A. Salam, A survey on subsurface signal propagation. Smart Cities **3**(4), 1513–1561 (2020)
11. U. Raza, A. Salam, Wireless underground communications in sewer and stormwater overflow monitoring: Radio waves through soil and asphalt medium. Information **11**(2), 98 (2020)
12. U. Raza, A. Salam, Zenneck waves in decision agriculture: An empirical verification and application in EM-based underground wireless power transfer. Smart Cities **3**(2), 308–340 (2020)
13. D. Roman, A. Braga, N. Shetty, P. Culligan, Design and modeling of an adaptively controlled rainwater harvesting system. Water **9**(12), 974 (2017)
14. A. Salam, M.C. Vuran, X. Dong, C. Argyropoulos, S. Irmak, A theoretical model of underground dipole antennas for communications in Internet of Underground Things. IEEE Trans. Antennas Propagat. **67**(6), 3996–4009 (2019)

15. A. Salam, M.C. Vuran, S. Irmak, A statistical impulse response model based on empirical characterization of wireless underground channel. IEEE Trans. Wireless Commun. **19**, 5966–5981 (2020)
16. A. Salam, M.C. Vuran, Impacts of soil type and moisture on the capacity of multi-carrier modulation in Internet of Underground Things, in *Proc. of the 25th ICCCN 2016*, Waikoloa, Hawaii, USA, Aug 2016
17. A. Salam, Pulses in the sand: Long range and high data rate communication techniques for next generation wireless underground networks. ETD Collection for University of Nebraska - Lincoln (AAI10826112) (2018)
18. A. Salam, A comparison of path loss variations in soil using planar and dipole antennas, in *2019 IEEE International Symposium on Antennas and Propagation* (IEEE, Jul 2019)
19. A. Salam, Design of subsurface phased array antennas for digital agriculture applications, in *Proc. 2019 IEEE International Symposium on Phased Array Systems and Technology (IEEE Array 2019)*, Waltham, MA, USA, October 2019
20. A. Salam, A path loss model for through the soil wireless communications in digital agriculture, in *2019 IEEE International Symposium on Antennas and Propagation* (IEEE, Jul 2019)
21. A. Salam, Sensor-free underground soil sensing. in *ASA, CSSA and SSSA International Annual Meetings (2019)* (ASA-CSSA-SSSA, 2019)
22. A. Salam, Subsurface MIMO: A beamforming design in Internet of Underground Things for digital agriculture applications. J. Sensor Actuat. Netw. **8**(3), 41 (2019)
23. A. Salam, *Underground Environment Aware MIMO Design Using Transmit and Receive Beamforming in Internet of Underground Things* (Springer International Publishing, Cham, 2019), pp. 1–15
24. A. Salam, An underground radio wave propagation prediction model for digital agriculture. Information **10**(4), 147 (2019)
25. A. Salam, Underground soil sensing using subsurface radio wave propagation, in *5th Global Workshop on Proximal Soil Sensing*, COLUMBIA, MO, May 2019
26. A. Salam, *Internet of Things for Environmental Sustainability and Climate Change* (Springer International Publishing, Cham, 2020), pp. 33–69
27. A. Salam, *Internet of Things for Sustainability: Perspectives in Privacy, Cybersecurity, and Future Trends* (Springer International Publishing, Cham, 2020), pp. 299–327
28. A. Salam, *Internet of Things for Sustainable Community Development*, 1 edn. (Springer Nature, New York, 2020)
29. A. Salam, *Internet of Things for Sustainable Community Development: Introduction and Overview* (Springer International Publishing, Cham, 2020), pp. 1–31
30. A. Salam, *Internet of Things for Sustainable Forestry* (Springer International Publishing, Cham, 2020), pp. 147–181
31. A. Salam, *Internet of Things for Sustainable Human Health* (Springer International Publishing, Cham, 2020), pp. 217–242
32. A. Salam, *Internet of Things for Sustainable Mining* (Springer International Publishing, Cham, 2020), pp. 243–271
33. A. Salam, *Internet of Things for Water Sustainability* (Springer International Publishing, Cham, 2020), pp. 113–145
34. A. Salam, *Internet of Things in Agricultural Innovation and Security* (Springer International Publishing, Cham, 2020), pp. 71–112
35. A. Salam, *Internet of Things in Sustainable Energy Systems* (Springer International Publishing, Cham, 2020), pp. 183–216
36. A. Salam, *Internet of Things in Water Management and Treatment* (Springer International Publishing, Cham, 2020), pp. 273–298
37. A. Salam, A.D. Hoang, A. Meghna, D.R. Martin, G. Guzman, Y.H. Yoon, J. Carlson, J. Kramer, K. Yansi, M. Kelly, et al., The future of emerging IoT paradigms: Architectures and technologies (2019)

38. A. Salam, U. Karabiyik, A cooperative overlay approach at the physical layer of cognitive radio for digital agriculture, in *Third International Balkan Conference on Communications and Networking 2019 (BalkanCom'19)*, Skopje, Macedonia, the former Yugoslav Republic of, June 2019
39. A. Salam, U. Raza, *Autonomous Irrigation Management in Decision Agriculture* (Springer International Publishing, Cham, 2020), pp. 379–398
40. A. Salam, U. Raza, *Current Advances in Internet of Underground Things* (Springer International Publishing, Cham, 2020), pp. 321–356
41. A. Salam, U. Raza, *Decision Agriculture* (Springer International Publishing, Cham, 2020), pp. 357–378
42. A. Salam, U. Raza, *Electromagnetic Characteristics of the Soil* (Springer International Publishing, Cham, 2020), pp. 39–59
43. A. Salam, U. Raza, *Modulation Schemes and Connectivity in Wireless Underground Channel* (Springer International Publishing, Cham, 2020), pp. 125–166
44. A. Salam, U. Raza, On burial depth of underground antenna in soil horizons for decision agriculture, in *2020 International Conference on Internet of Things (ICIOT-2020)*, Honolulu, USA, September 2020
45. A. Salam, U. Raza, *Signals in the Soil*1 edn. (Springer Nature, New York, 2020)
46. A. Salam, U. Raza, *Signals in the Soil: An Introduction to Wireless Underground Communications* (Springer International Publishing, Cham, 2020), pp. 3–38
47. A. Salam, U. Raza, *Signals in the Soil: Subsurface Sensing* (Springer International Publishing, Cham, 2020), pp. 251–297
48. A. Salam, U. Raza, *Signals in the Soil: Underground Antennas* (Springer International Publishing, Cham, 2020), pp. 189–215
49. A. Salam, U. Raza, *Soil Moisture and Permittivity Estimation* (Springer International Publishing, Cham, 2020), pp. 299–317
50. A. Salam, U. Raza, *Underground Phased Arrays and Beamforming Applications* (Springer International Publishing, Cham, 2020), pp. 217–248
51. A. Salam, U. Raza, *Underground Wireless Channel Bandwidth and Capacity* (Springer International Publishing, Cham, 2020), pp. 167–188
52. A. Salam, U. Raza, *Variable Rate Applications in Decision Agriculture* (Springer International Publishing, Cham, 2020), pp. 399–423.
53. A. Salam, U. Raza, *Wireless Underground Channel Modeling* (Springer International Publishing, Cham, 2020), pp. 61–121
54. A. Salam, S. Shah, Internet of Things in smart agriculture: Enabling technologies, in *2019 IEEE 5th World Forum on Internet of Things (WF-IoT) (WF-IoT 2019)*, Limerick, Ireland, April 2019
55. A. Salam, M.C. Vuran, EM-based wireless underground sensor networks, in *Underground Sensing*, ed. by S. Pamukcu, L. Cheng (Academic Press, Cambridge, 2018), pp. 247–285
56. A. Salam, M.C. Vuran, S. Irmak, Di-Sense: In situ real-time permittivity estimation and soil moisture sensing using wireless underground communications. Comput. Netw. **151**, 31–41 (2019)
57. A. Salam, M.C. Vuran, Smart underground antenna arrays: A soil moisture adaptive beamforming approach, in *Proc. IEEE INFOCOM 2017*, Atlanta, USA, May 2017
58. A. Salam, M.C. Vuran, Wireless underground channel diversity reception with multiple antennas for Internet of Underground Things, in *Proc. IEEE ICC 2017*, Paris, France, May 2017
59. A. Salam, M.C. Vuran, S. Irmak, Pulses in the sand: Impulse response analysis of wireless underground channel, in *The 35th Annual IEEE International Conference on Computer Communications (INFOCOM 2016)*, San Francisco, USA, April 2016
60. A. Salam, M.C. Vuran, S. Irmak, Towards Internet of Underground Things in smart lighting: A statistical model of wireless underground channel, in *Proc. 14th IEEE International Conference on Networking, Sensing and Control (IEEE ICNSC)*, Calabria, Italy, May 2017

61. S. Temel, M.C. Vuran, M.M.R. Lunar, Z. Zhao, A. Salam, R.K. Faller, C. Stolle, Vehicle-to-barrier communication during real-world vehicle crash tests. Comput. Commun. **127**, 172–186 (2018)
62. M.C. Vuran, A. Salam, R. Wong, S. Irmak, Internet of Underground Things in precision agriculture: Architecture and technology aspects. Ad Hoc Netw. (2018). https://doi.org/10.1016/j.adhoc.2018.07.017
63. M.C. Vuran, A. Salam, R. Wong, S. Irmak, Internet of Underground Things: Sensing and communications on the field for precision agriculture, in *2018 IEEE 4th World Forum on Internet of Things (WF-IoT) (WF-IoT 2018)*, Singapore, February 2018
64. J. Wright, D. Marchese, Briefing: Continuous monitoring and adaptive control: the smartstorm water management solution. Proc. Inst. Civil Eng. Smart Infrastruct. Construct. **170**(4), 86–89 (2018)

Chapter 3
Smart Sewer Soil Modeling for IoT Communications

Check for updates

3.1 Introduction

This investigation addresses the soil penetrability considering soil parameters (type and moisture) and signal frequency. Experiments are performed to measure the skin depth using two types of soils: clay and gravel [5–19]. Moisture content for these types are **10**%, **50**%, and **90**%. Finally, signal frequency is used between the range of **0.5**–4.5 GHz. Test samples of surface area **1.22** mm × **1.22** mm were compounded and sealed in a polyethylene to retain overall soil moisture content and soil moisture content within the layers.

3.2 Theoretical Model

Attenuation and phase shift happen to all the waves traveling through the complex medium. In this section, theoretical model is proposed by outlining the properties of complex medium. Equation (7.1) calculates the propagation constant of plane waves as [20–26]:

$$\gamma = j\omega\sqrt{\mu\epsilon}\left[1 - j\frac{\sigma}{\omega\epsilon}\right]^{\frac{1}{2}}. \tag{3.1}$$

Equation (7.1) can be rewritten as

$$\gamma = \alpha(\omega) + j\beta(\omega), \tag{3.2}$$

where $\alpha(\omega)$ and $\beta(\omega)$ can be calculated as follows:

A. Salam, *Internet of Things in Smart Sewer and Drainage Systems*,
https://doi.org/10.1007/978-3-031-48109-3_3

$$\alpha(\omega) = \left\{ \frac{\omega^2 \mu \epsilon}{2} \left[\sqrt{1 + \left(\frac{\sigma}{\omega \epsilon} \right)^2} - 1 \right] \right\}^{\frac{1}{2}} \tag{3.3}$$

$$\beta(\omega) = \left\{ \frac{\omega^2 \mu \epsilon}{2} \left[\sqrt{1 + \left(\frac{\sigma}{\omega \epsilon} \right)^2} + 1 \right] \right\}^{\frac{1}{2}}, \tag{3.4}$$

where $\left(\frac{\sigma}{\omega\epsilon} << 1\right)$ is negligible for small losses; therefore, $\alpha(\omega)$ and $\beta(\omega)$ are to be reduced to

$$\alpha(\omega) \approx \frac{\sigma}{2} \sqrt{\frac{\mu}{\epsilon}} \tag{3.5}$$

$$\beta(\omega) \approx \frac{\sigma}{2} \sqrt{\mu \epsilon}. \tag{3.6}$$

Intrinsic impedance of the medium is calculated as [27–38]

$$\eta = \sqrt{\frac{\mu}{\epsilon \left(1 - j \frac{\sigma}{\omega \epsilon}\right)}}. \tag{3.7}$$

The medium used for this model is an isotropic and consists of layers of air and aluminum and a layer of finite thickness in between the former two layers. Moreover, it treats the layer as a non-magnetic layer, i.e., $\mu = \mu_0$, where $\mu_0 = 4 \times 10^7$ H/m, and considers only electrical properties of the layer. Electrical properties under consideration include permittivity (ϵ), permeability (μ), and conductivity (σ), where ϵ is not dependent on frequency.

The model assumes ideal horns due to which only reflected waves are received. However, in practice, a lot of design considerations are required for the reduction of direct coupling. The horns used were not ideal; however, model of single ray normal incidence is used (Figs. 3.1 and 3.2).

Reflection coefficient in medium 1 (see [1]) can be written as

$$\Gamma_{in} = \frac{\Gamma_1 - e^{-2\gamma_2 l_2}}{1 - \Gamma e^{-2\gamma_2 l_2}}, \tag{3.8}$$

where Γ_1 represents the reflection coefficient of the medium and is calculated as

$$\Gamma_1 = \frac{1 - \sqrt{\epsilon_2'}}{1 + \sqrt{\epsilon_2'}}, \tag{3.9}$$

and γ_2 is calculated by Eqs. (3.2), (3.3), and (3.4). Magnitude of Γ_{in} will be calculated as function of dielectric constant and loss, and frequency and layer

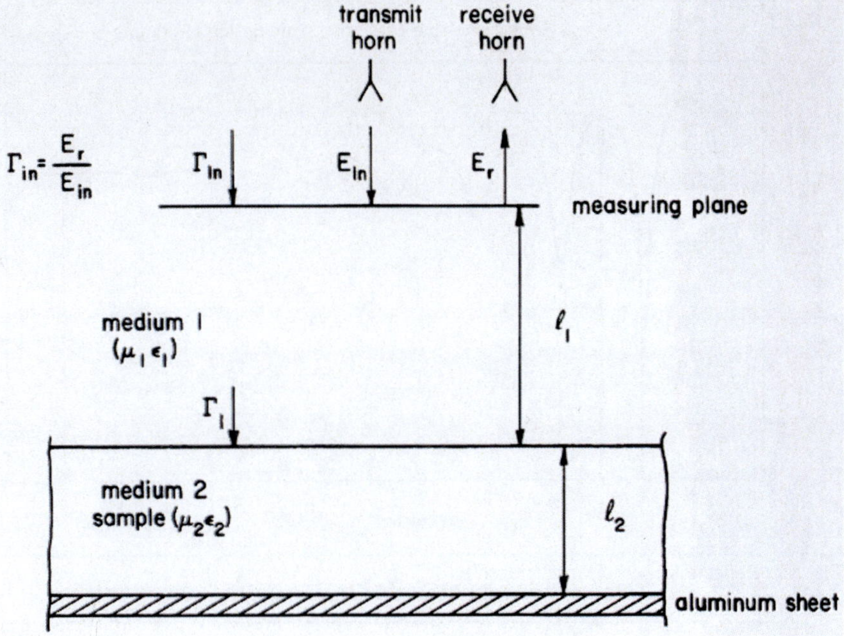

Fig. 3.1 Plane waves normally incident on a plane dielectric boundary

thickness. There will be a minima or maxima every 2π radians. Therefore, on varying ω, occurrence of successive minima will be as follows [31–37]:

$$2(\Delta\beta_2)l_2 = 2\pi \tag{3.10}$$

from which

$$\sqrt{\epsilon_2'} = \frac{c}{2l_2(f_2 - f_1)}, \tag{3.11}$$

where f_1 and f_2 are the frequency point for the occurrence of successive minima.

It has already been investigated in [2] that maxima of Γ_{in} can be approximated at $\cos 2\beta_2 l_2 = 1$. Therefore, Eq. (3.8) can be rewritten as [38–48]

$$\Gamma_m = \frac{\Gamma_1 - e^{-2\alpha_2 l_2}}{1 - \Gamma e^{-2\alpha_2 l_2}}, \tag{3.12}$$

where attenuator factor $-2\alpha_2 l_2$ is calculated as [49–51]

$$-2\alpha_2 l_2 = ln\left[\frac{\Gamma_1 - \Gamma_m}{1 - \Gamma_1\Gamma_m}\right]. \tag{3.13}$$

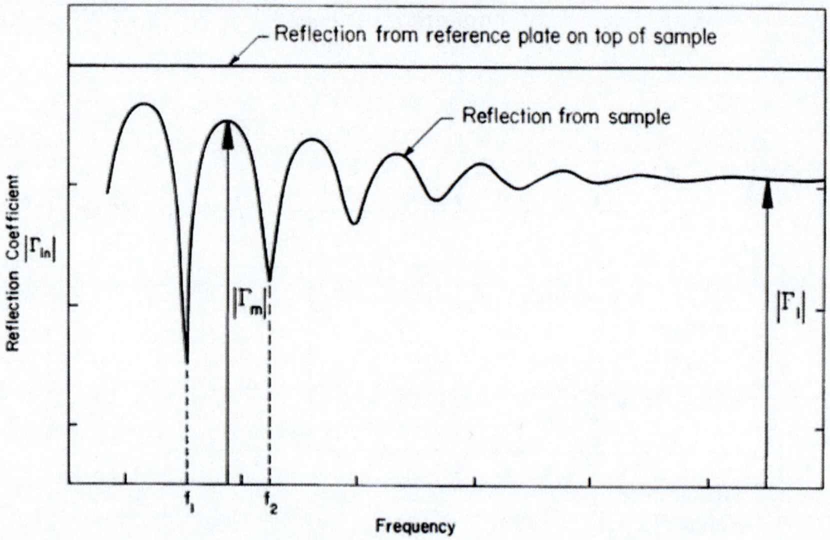

Fig. 3.2 Reflection coefficient relationship

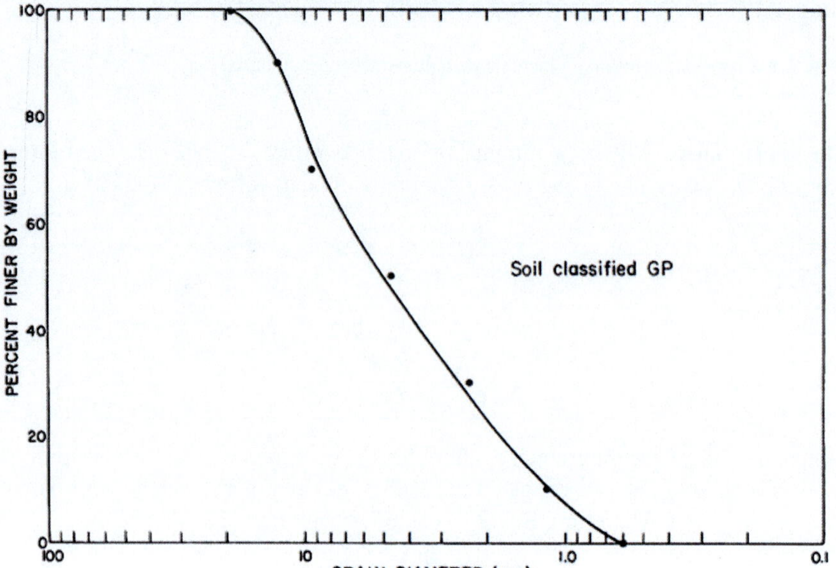

Fig. 3.3 Grain size distribution of gravel sample

Since Γ_1 in Eq. (3.9) will always be less than 0, therefore, Γ_m will also be negative (Fig. 3.3).

Figure 4.2 relates both reflection coefficients $|\Gamma_1|$ and $|\Gamma_{in}|$. $|\Gamma_1|$ is deduced from Eq. (3.11), and $|\Gamma_{in}|$ is measured using technique of return loss measurement in [3].

The thickness of the sample is uniform; therefore, attenuation can be calculated from Eq. (3.13).

Skin depth is a depth at which signal is attenuated by 1/e of its original strength, and it is calculated as

$$\alpha_2 \delta = 1, \tag{3.14}$$

where δ represents the skin depth [52, 53].

3.2.1 *Gravel*

Figure 4.3 shows the grain curve from grinding gravel with the mixture of coarse and fine concrete. The resultant soil mixture had rating of poorly graded gravel (GP) as per the system of Unified Soil Classification (see [4]). A modified Proctor method is used to characterize soil compaction, and the results are shown in Fig. 3.4. These results help in selecting the soil moisture values for soil compaction in testbed to get the saturation values of **10**%. Actual saturation can only be predicted once the testbed is prepared [54, 55].

A mixture of dry gravel and water was prepared in concrete mixer and compacted through a combination of gasoline-powered vibratory compactor and hand compaction. After the compaction process, the box is weighed to calculate the different parameters of soil in the box such as moisture content, dry density, grain average specific gravity, void ratio, and degree of saturation.

Two soil beds were made, and box dimensions, for each bed, were kept the same, i.e., **1.22** m × **1.22** m × **20.3** cm. It was important to prevent the soil moisture loss from the testbed; therefore, boxes were lined with the sheets of polyethylene plastic.

The above-mentioned procedure was used to prepare a testbed with **10**% saturation. The degree of saturation was increased to **50**% after taking the measurement. For this purpose, an adequate amount of water was calculated from the dry density of the prepared soil, and the water was introduced through the surface. Box was weighed again to determine new degree of saturation, therefore keeping the value of dry density of soil constant between both specimens. Same procedure was repeated to achieve **90**% of saturation; however, this time it was very difficult due to trapped air present in the soil. Table 3.1 shows the actual values of data on both soil testbeds with all three degrees of saturation level.

3.2.2 *Clay*

Major difficulty was to find the large quantity of soil having a CL classification. After selecting the soil, it was crushed, dried, and sieved. Figure 3.5 shows the grain size distribution curve. This curve also helps in finding Atterberg limits.

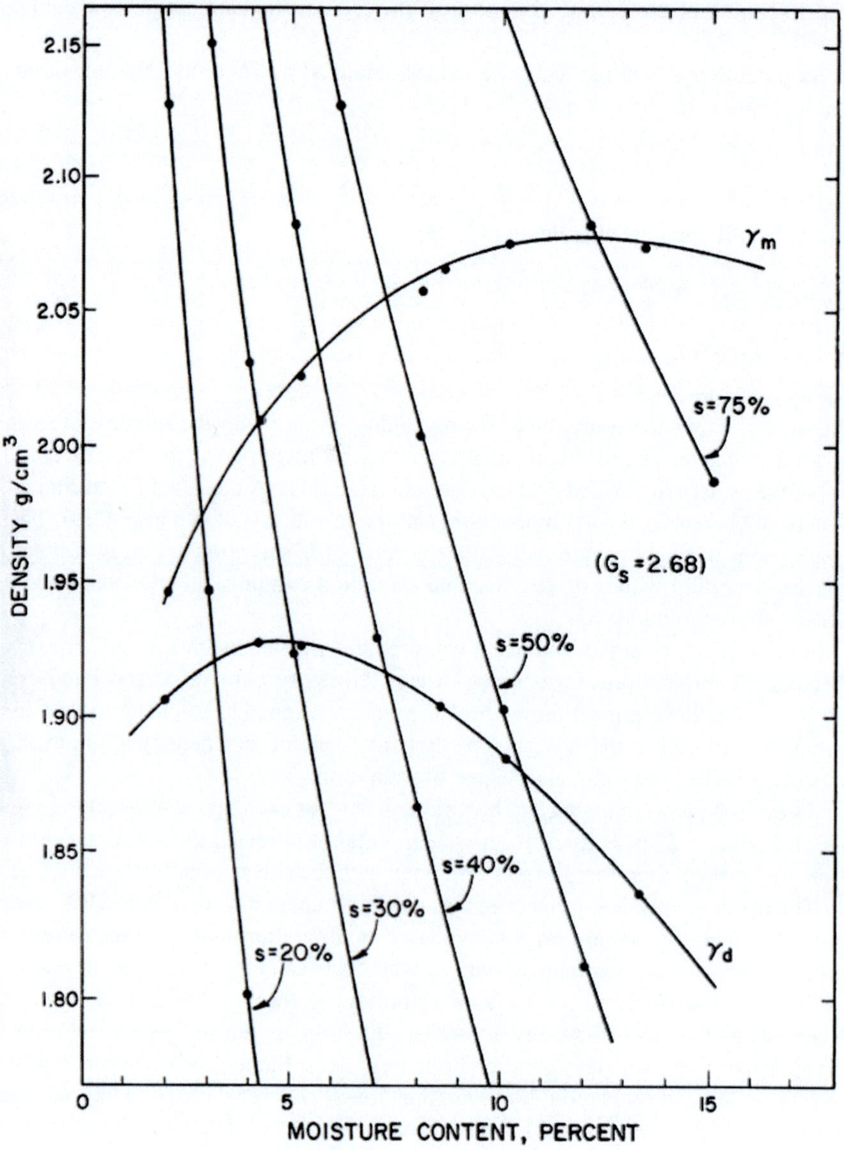

Fig. 3.4 Modified Proctor moisture and density curves of gravel

The resultant soil mixture had rating of CL as per the system of Unified Soil Classification (see [4]). Figure 4.4 shows the compaction characteristics of the sample.

A mixture of dry soil and water was prepared in soil mixer in batches of **110** kg and compacted with a hand compactor. The hand compactor was used to achieve

Table 3.1 Properties of compacted soil

		Moisture content (%)	Dry density (g/cm^3)	Degree of saturation (%)
Gravel	Bed 1	2.0	1.802	11.4
		8.6	1.802	48.9
		15.1	1.802	85.9
	Bed 2	2.0	1.874	12.8
		7.4	1.874	47.3
		12.7	1.874	81.2
Clay		3.5	1.352	9.6
		11.0	1.528	39.3
		23.2	1.528	82.8

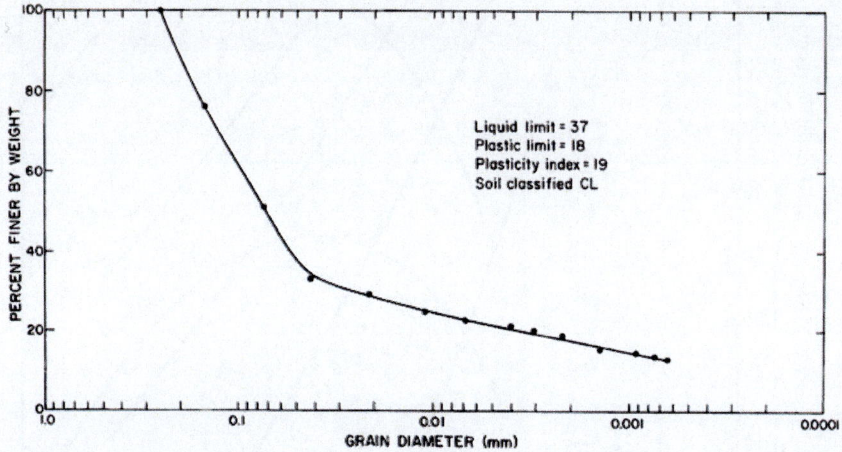

Fig. 3.5 Grain size distribution of clay sample

required density and therefore resulting in desired saturation level (Figs. 3.6 and 3.7). The degree of saturation was increased to **50%** after taking the measurement. For this purpose, additional water was mixed with the soil using a rammer-type compaction machine, powered by gasoline, to achieve higher density. Four plumbing fittings were fixed at the bottom of the box to add more water to it and achieve **90%** saturation. This was done doing compaction at **90%** saturation and retaining compaction from **50%** saturation. Table 3.2 shows the data for the clay sample.

3.3 Experimental Results and Discussion

Initial results show that polyethylene sealing was very helpful in retaining the soil moisture. Moreover, a uniformity test was done by comparative analysis of dielectric constant of surface and layer, and it was observed that soil moisture was retained

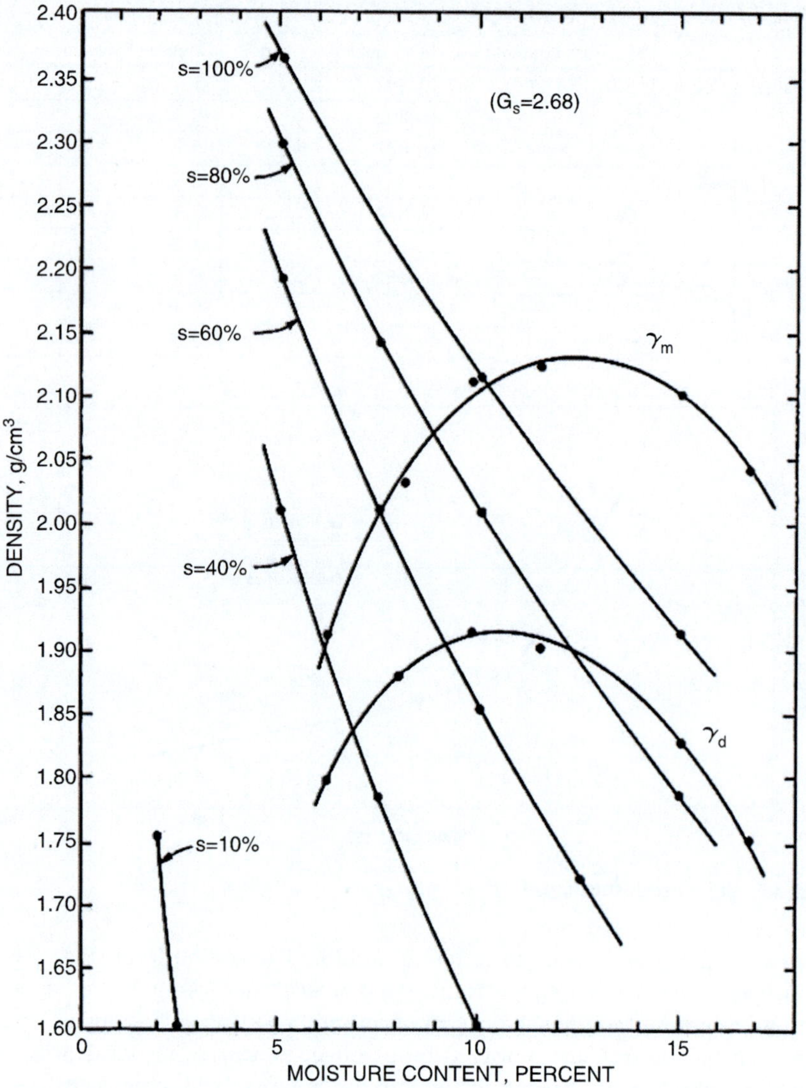

Fig. 3.6 Modified proctor moisture and density curves of clay

uniformly throughout the sample. The surface area and thickness for all initial samples were **1.22** m × **1.22** m and **20.3** cm. However, these values were reduced for the samples with higher moisture content so that measurement can be done easily for higher microwave absorption rate. Figure 4.5 shows the sample of measured from these experiments.

Skin depth was calculated form the measured data and is shown in Table 3.2. In addition to skin depth, Table 3.2 also lists different important parameters for

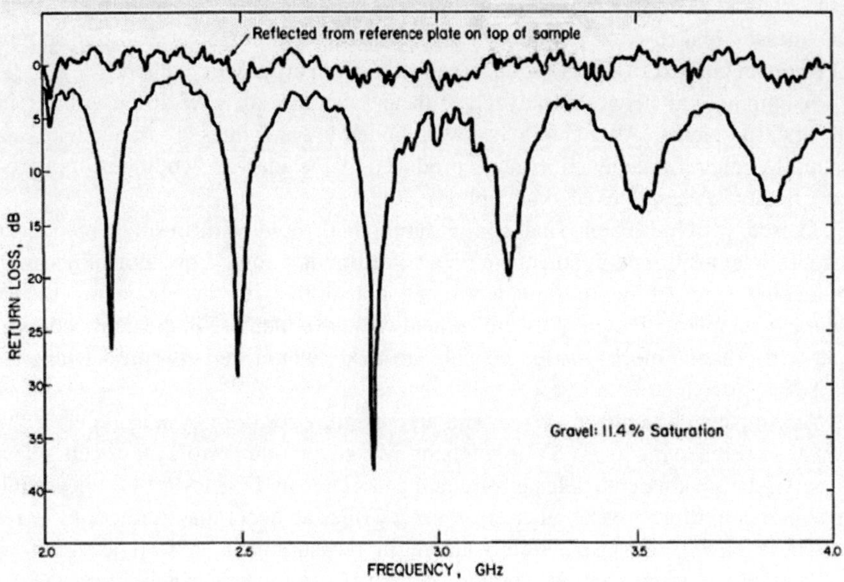

Fig. 3.7 Samples of measured data

Table 3.2 Skin depths

Material	Saturation (%)	Sample thickness (cm)	Frequency (GHz)	Relative dielectric constant	Degree of saturation (%)
Clay	9.6	20.2	0.5–1.0	4.0	30–42
	9.6	20.2	2.3–2.6	5.1	13–19
	9.6	10.8	4.3–4.5	4.8	7–15
	39.3	8.9	0.5–1.0	9.2	10–20
	39.3	8.9	2.3–2.6	7.7	6–18
	39.3	8.9	4.0–4.5	7.5	6–9
	82.8	2.5	0.6–1.7	26.2	5–20
	82.8	2.5	2.0–3.0	22.1	3–6
	82.8	2.5	3.0–4.5	20.8	1.5–3.5
Gravel	11.4	20.2	0.5–1.0	4.9	160–315
	11.4	20.2	2.2–2.5	4.9	60–80
	11.4	20.2	4.0–4.5	4.6	17–30
	48.9	20.2	0.5–1.0	11.2	30–200
	48.9	12.1	2.2–2.5	13.3	15–35
	48.9	7.6	4.0–4.5	14.7	4–8
	85.9	20.2	0.5–1.0	15.2	40–60
	85.9	20.2	2.2–2.5	15.5	5–9
	85.9	9.7	4.0–4.5	–	–

each sample. These parameters include measured relative dielectric constants, thicknesses, and relevant frequency ranges. Instead of doing error analysis, sources of errors in context of effect produced on skin depth are discussed here.

A commercial network analyzer, with an accuracy of ±**0.5** dB, was used for the measurements. Attenuation is lower under lower values of frequencies and saturation; therefore, any uncertainty produced by the analyzer would result in great uncertainty in the values of skin depth.

In order to achieve a physical configuration that would confirm single ray theory, sample was positioned in antenna's Fresnel diffraction zone. Low-moisture sample producing entertaining in skin depth will not align with the single ray theory, whereas sample with high moisture content will have increase in dielectric constant and will produce more focused beams. Error rate would lie between the range of **1**–2 dB.

Any mismatch between system and antenna may result in systematic error that can be lowered using low-VSWR antennas. Moreover, other errors may occur due to uncertainty of skin depth. This error could be as large as **1** dB for a **10%** uncertainty in skin depth. Moreover, another error, up to **2** dB, can occur due to fractional error.

These errors emphasize improvement in measurement as well as hardware to reduce uncertainty in skin depth. Moreover, more data can be gathered for improvement. Samples with varying percentages of soil moisture content can be prepared. Field experiments with in situ measurements can also be done to cater for the phenomenon that may not occur in lab test. Finally, different materials such as asphalt and concrete can also be used for the experiments.

3.4 Conclusion

The purpose of this investigation was to explore the relation between microwave penetration and soil properties (type and soil moisture content), and frequency. To that end, experiments were performed on different types of soil samples with soil moisture content values of **10%**, **50%**, and **90%** and a frequency range of **0.5**–4.5 GHz.

Samples were sealed in a polyethylene bag to retain a uniform soil moisture value for a long period of time. A state-of-the-art microwave equipment was used, which has the ability to penetrate five-skin-depth layer of the material at +10 dBm.

Signal penetration is dependent upon the moisture content of the material. If total reflection is considered, a signal with a frequency of **0.5** GHz can pass through 150 cm layer and 25 cm layer for saturation levels of 10% and 90%, respectively. Similarly, penetration are 35 cm and 1 cm for saturation levels of 10% and 90% clay, respectively, under the frequency of 4.5 GHz.

For gravel, a signal with a frequency of **0.5** GHz can pass through 800 cm layer and 120 cm layer for saturation levels of 10% and 90%, respectively. Hence, it was observed that gravel had more skin depth than clay.

The above-mentioned penetration levels were measured under the effect of surface reflection. These levels should be measured under no or minimum reflection. Some techniques to reduce reflection include adding an ancillary circuit, using a gating circuitry, or measurement of penetration in time domain instead of frequency domain. Moreover, penetration depth here is measured under the assumption of total reflection at interfaces of clay and gravel. Material with both clay and gravel will have reduced penetration ability because of no total reflection at interface.

Moisture content also plays an important role in affecting the penetration depths. The saturated gravel will have less penetration depth at as compared to the saturated clay at 10% saturation. It will further decrease when the moisture content is increased up to 90%. These measurements assume that surface reflection has no effect on measurement capability of the system.

References

1. A. Konda, A. Rau, M.A. Stoller, J.M. Taylor, A. Salam, G.A. Pribil, C. Argyropoulos, S.A. Morin, Soft microreactors for the deposition of conductive metallic traces on planar, embossed, and curved surfaces. Adv. Funct. Mater. **28**(40), 1803020 (2018)
2. U. Raza, A. Salam, On-site and external power transfer and energy harvesting in underground wireless. Electronics **9**(4), 681 (2020)
3. U. Raza, A. Salam, A survey on subsurface signal propagation. Smart Cities **3**(4), 1513–1561 (2020)
4. U. Raza, A. Salam, Wireless underground communications in sewer and stormwater overflow monitoring: Radio waves through soil and asphalt medium. Information **11**(2), 98 (2020)
5. U. Raza, A. Salam, Zenneck waves in decision agriculture: An empirical verification and application in EM-based underground wireless power transfer. Smart Cities **3**(2), 308–340 (2020)
6. A. Salam, Pulses in the sand: Long range and high data rate communication techniques for next generation wireless underground networks. *ETD Collection for University of Nebraska - Lincoln* (AAI10826112, 2018)
7. A. Salam, A comparison of path loss variations in soil using planar and dipole antennas, in *2019 IEEE International Symposium on Antennas and Propagation* (IEEE, Jul 2019)
8. A. Salam, Design of subsurface phased array antennas for digital agriculture applications, in *Proc. 2019 IEEE International Symposium on Phased Array Systems and Technology (IEEE Array 2019)*, Waltham, MA, USA, October 2019
9. A. Salam, A path loss model for through the soil wireless communications in digital agriculture, in *2019 IEEE International Symposium on Antennas and Propagation* (IEEE, Jul 2019)
10. A. Salam, Sensor-free underground soil sensing, in *ASA, CSSA and SSSA International Annual Meetings (2019)* (ASA-CSSA-SSSA, 2019)
11. A. Salam, Subsurface MIMO: A beamforming design in Internet of Underground Things for digital agriculture applications. J. Sensor Actuat. Netw. **8**(3), 41 (2019)
12. A. Salam, *Underground Environment Aware MIMO Design Using Transmit and Receive Beamforming in Internet of Underground Things* (Springer International Publishing, Cham, 2019), pp. 1–15
13. A. Salam, An underground radio wave propagation prediction model for digital agriculture. Information **10**(4), 147 (2019)

14. A. Salam, Underground soil sensing using subsurface radio wave propagation, in *5th Global Workshop on Proximal Soil Sensing*, COLUMBIA, MO, May 2019
15. A. Salam, *Internet of Things for Environmental Sustainability and Climate Change* (Springer International Publishing, Cham, 2020), pp. 33–69
16. A. Salam, *Internet of Things for Sustainability: Perspectives in Privacy, Cybersecurity, and Future Trends* (Springer International Publishing, Cham, 2020), pp. 299–327
17. A. Salam, *Internet of Things for Sustainable Community Development*, 1st edn. (Springer Nature, New York, 2020)
18. A. Salam, *Internet of Things for Sustainable Community Development: Introduction and Overview* (Springer International Publishing, Cham, 2020), pp. 1–31
19. A. Salam, *Internet of Things for Sustainable Forestry* (Springer International Publishing, Cham, 2020), pp. 147–181
20. A. Salam, *Internet of Things for Sustainable Human Health* (Springer International Publishing, Cham, 2020), pp. 217–242
21. A. Salam, *Internet of Things for Sustainable Mining* (Springer International Publishing, Cham, 2020), pp. 243–271
22. A. Salam, *Internet of Things for Water Sustainability* (Springer International Publishing, Cham, 2020), pp. 113–145
23. A. Salam, *Internet of Things in Agricultural Innovation and Security* (Springer International Publishing, Cham, 2020), pp. 71–112
24. A. Salam, *Internet of Things in Sustainable Energy Systems* (Springer International Publishing, Cham, 2020), pp. 183–216
25. A. Salam, *Internet of Things in Water Management and Treatment* (Springer International Publishing, Cham, 2020), pp. 273–298
26. A. Salam, A.D. Hoang, A. Meghna, D.R. Martin, G. Guzman, Y.H. Yoon, J. Carlson, J. Kramer, K. Yansi, M. Kelly, et al., The future of emerging IoT paradigms: Architectures and technologies (2019)
27. A. Salam, U. Karabiyik, A cooperative overlay approach at the physical layer of cognitive radio for digital agriculture, in *Third International Balkan Conference on Communications and Networking 2019 (BalkanCom'19)*, Skopje, Macedonia, the former Yugoslav Republic of, June 2019
28. A. Salam, M.C. Vuran, Impacts of soil type and moisture on the capacity of multi-carrier modulation in Internet of Underground Things, in *Proc. of the 25th ICCCN 2016*, Waikoloa, Hawaii, USA, Aug 2016
29. A. Salam, M.C. Vuran, X. Dong, C. Argyropoulos, S. Irmak, A theoretical model of underground dipole antennas for communications in Internet of Underground Things. IEEE Trans. Antenn. Propagat. **67**(6), 3996–4009 (2019)
30. A. Salam, M.C. Vuran, S. Irmak, A statistical impulse response model based on empirical characterization of wireless underground channel. IEEE Trans. Wirel. Commun. **19**, 5966–5981 (2020)
31. A. Salam, U. Raza, *Autonomous Irrigation Management in Decision Agriculture* (Springer International Publishing, Cham, 2020), pp. 379–398
32. A. Salam, U. Raza, *Current Advances in Internet of Underground Things* (Springer International Publishing, Cham, 2020), pp. 321–356
33. A. Salam, U. Raza, *Decision Agriculture* (Springer International Publishing, Cham, 2020), pp. 357–378
34. A. Salam, U. Raza, *Electromagnetic Characteristics of the Soil* (Springer International Publishing, Cham, 2020), pp. 39–59
35. A. Salam, U. Raza, *Modulation Schemes and Connectivity in Wireless Underground Channel* (Springer International Publishing, Cham, 2020), pp. 125–166
36. A. Salam, U. Raza, On burial depth of underground antenna in soil horizons for decision agriculture, in *2020 International Conference on Internet of Things (ICIOT-2020)*, Honolulu, USA, September 2020
37. A. Salam, U. Raza, *Signals in the Soil*, 1st edn. (Springer Nature, New York, 2020)

38. A. Salam, U. Raza, *Signals in the Soil: An Introduction to Wireless Underground Communications* (Springer International Publishing, Cham, 2020), pp. 3–38
39. A. Salam, U. Raza, *Signals in the Soil: Subsurface Sensing* (Springer International Publishing, Cham, 2020), pp. 251–297
40. A. Salam, U. Raza, *Signals in the Soil: Underground Antennas* (Springer International Publishing, Cham, 2020), pp. 189–215
41. A. Salam, U. Raza, *Soil Moisture and Permittivity Estimation* (Springer International Publishing, Cham, 2020), pp. 299–317
42. A. Salam, U. Raza, *Underground Phased Arrays and Beamforming Applications* (Springer International Publishing, Cham, 2020), pp. 217–248
43. A. Salam, U. Raza, *Underground Wireless Channel Bandwidth and Capacity* (Springer International Publishing, Cham, 2020), pp. 167–188
44. A. Salam, U. Raza, *Variable Rate Applications in Decision Agriculture* (Springer International Publishing, Cham, 2020), pp. 399–423
45. A. Salam, U. Raza, *Wireless Underground Channel Modeling* (Springer International Publishing, Cham, 2020), pp. 61–121
46. A. Salam, S. Shah, Internet of Things in smart agriculture: Enabling technologies, in *2019 IEEE 5th World Forum on Internet of Things (WF-IoT) (WF-IoT 2019)*, Limerick, Ireland, April 2019
47. A. Salam, M.C. Vuran, EM-based wireless underground sensor networks, in *Underground Sensing*, ed. by S. Pamukcu, L. Cheng (Academic Press, Cambridge, 2018), pp. 247–285
48. A. Salam, M.C. Vuran, Smart underground antenna arrays: A soil moisture adaptive beamforming approach, in *Proc. IEEE INFOCOM 2017*, Atlanta, USA, May 2017
49. A. Salam, M.C. Vuran, Wireless underground channel diversity reception with multiple antennas for Internet of Underground Things, in *Proc. IEEE ICC 2017*, Paris, France, May 2017
50. A. Salam, M.C. Vuran, S. Irmak, Pulses in the sand: Impulse response analysis of wireless underground channel, in *The 35th Annual IEEE International Conference on Computer Communications (INFOCOM 2016)*, San Francisco, USA, April 2016
51. A. Salam, M.C. Vuran, S. Irmak, Towards Internet of Underground Things in smart lighting: A statistical model of wireless underground channel, in *Proc. 14th IEEE International Conference on Networking, Sensing and Control (IEEE ICNSC)*, Calabria, Italy, May 2017
52. A. Salam, M.C. Vuran, S. Irmak, Di-Sense: In situ real-time permittivity estimation and soil moisture sensing using wireless underground communications. Comput. Networks **151**, 31–41 (2019)
53. S. Temel, M.C. Vuran, M.M.R. Lunar, Z. Zhao, A. Salam, R.K. Faller, C. Stolle, Vehicle-to-barrier communication during real-world vehicle crash tests. Comput. Commun. **127**, 172–186 (2018)
54. M.C. Vuran, A. Salam, R. Wong, S. Irmak, Internet of Underground Things in precision agriculture: Architecture and technology aspects. Ad Hoc Networks (2018). https://doi.org/10.1016/j.adhoc.2018.07.017
55. M.C. Vuran, A. Salam, R. Wong, S. Irmak, Internet of Underground Things: Sensing and communications on the field for precision agriculture, in *2018 IEEE 4th World Forum on Internet of Things (WF-IoT) (WF-IoT 2018)*, Singapore, February 2018

Chapter 4
City Scale Drainage System IoT Testbeds

4.1 Introduction

Advancement in Internet of Things (IoT) applications is also resulting in evolution of wireless standards. Advanced IoT applications have new requirements, and to address those, new wireless standards are being developed. Seamless underground and over-the-air wireless communication is the demand of diverse IoT applications, e.g., smart agriculture and infrastructure. With such a diverse set of IoT applications and wireless technologies, it is very important to have a diverse set of tools to perform experiments. This could be difficult and expensive and not every researcher may be able to build these tools. Therefore, open testbeds [8, 9, 13, 14, 17, 19, 81] can provide the researcher an experiment platform to emulate these technologies and come up with new creative methodologies. It is important to have testbeds that mimic real environment and configurations with reusable datasets for advancing the research in RFML technology.

Considering the above-mentioned requirements, a testbed, Cognitive Secure Cloud-Radio Access Network (CoSeC-RAN), is developed with the cooperation of Lincoln City. This testbed covers a radius of 1.1 square miles spanned over two campuses of University of Nebraska-Lincoln. Other features of this testbed include a Cloud-Radio Access Network (C-RAN), five software-defined distribution units (SD-DUs), five sub-6 GHz Software-Defined Radios (SDRs), and a 4×4 MIMO antenna array. The array is connected to cloud through a 20 giga bit per second network. Moreover, in order to assess underground (UG) communication environment, UG MIMO antennas are also installed along with the SD-DUs. To address high computational requirements, CoSeC-RAN is equipped with an array of graphics processing units (GPUs) and Central Processing Units (CPUs) provided by the Holland Computing Center (HCC) [8]. HCC is a very high performance computing cloud. Finally a Field-Programmable Gate Array (FPGA) is also used at edges, i.e., SD-DUs and front-haul network.

A. Salam, *Internet of Things in Smart Sewer and Drainage Systems*,
https://doi.org/10.1007/978-3-031-48109-3_4

4.2 Related Work

4.2.1 Academic Testbeds

There have been many testbeds proposed by different academic institutions, for example, CORNET [14] and LTE-CORNET [13] from Virginia Tech, ORBIT [19] from Rutgers, and WARP [11] from Rice [20–28].

The orbit is made up of various indoor and outdoor nodes spread over the whole campus of Rutgers University. The orbit uses multiple Wi-Fi, WiMAX, Zigbee, and SDR radio nodes. The indoor radio grid is made up of 400 nodes, whereas outdoor nodes are deployed on vehicles which roam around the campus [5]. WARP helps researchers to build their own testbed using its hardware- and software-based reference designs. CORNET [13, 14] also spreads over the whole university and uses both hardware and software components. Hardware includes 48 SDR transceivers, e.g., USRP [5] and WARP [11]. Software components include GNURadio [10], LabView, and MATLAB [6].

4.2.2 Large-Scale Industrial Testbeds

An open testbed, FIT [81], is built by France on a very large scale. FIT has three major components: FIT-Wireless, FIT IoT-LAB, and FIT Cloud. FIT-Wireless consists of Wi-Fi, 5G, and cognitive radios. FIT IoT-LAB testbed consists of 2728 nodes over 6 different sites. FIT-Cloud is a cloud platform that combines working of FIT-Wireless and FIT IoT-LAB [13].

Platforms for Advanced Wireless Research (PAWR) [18] is an initiative of National Science Foundation in the USA. Under this initiative, three testbeds were developed by three different universities. Each testbed is constructed with cooperation with multiple universities focusing on different technologies. Table 4.1 shows the detail on all testbeds developed under this program [76, 77].

4.2.3 CoSeC-RAN Testbed

CoSeC-RAN testbed [81] focuses more on outdoor real-world settings, and however, it also shares some features from the above-mentioned testbeds. The commonality includes hardware and software from ORBIT and CORNET, Cloud-Radio Access Network (CRAN) similar to POWDERRENEW, and COSMOS [2, 17]. Other features of CoSeC-RAN include FPGAs, CPUs, GPUs, sub-1 GHz frequency, and support for underground wireless research with an ability to be extended to mmWave radio communication.

Table 4.1 PAWR testbeds

Testbed	Partner University and Organizations	Covered area	Technology
POWDERRENEW [6]	– University of Utah – Rice University	3.3 square miles	Dynamic spectrum sharing and massive MIMO capabilities
COSMOS [7]	– Rutgers University – Columbia University – New York University	1 square miles	Millimeter-wave (mmWave) Radio communications and Dynamic optical switching technologies
AERPAW [14]	– North Carolina State University – Mississippi State University – North Carolina Department of Transportation – Purdue University – University of South Carolina	–	Advanced wireless features for Unmanned Aircraft Systems (UAS) integration of and UAS into the national air-space

4.3 Testbed Architecture

This section discusses three major components of CoSeC-RAN testbed shown in Fig. 4.1.

4.3.1 Hardware

4.3.1.1 Basic Features

CoSeC-RAN is composed of multiple hardware components: Software-Defined Distribution Units (SD-DUs), cloud-based central unit, and fiber Ethernet-based fronthaul network. SD-DU consists of MIMO SDRs which has the capability of being tuned up to sub-6 GHz frequency. The current deployment has a total of five SD-DUs deployed: four on campus and one at street. These SD-DUs are connected to HCC using a fronthaul Ethernet network. HCC, due to its high performance capabilities, is used for data processing and storage [29–37].

4.3.1.2 Experiments

This architecture with 4×4 MIMO SDR transceiver provides an opportunity to conduct experiments on multiple levels, e.g., mesh network, point-to-point MIMO

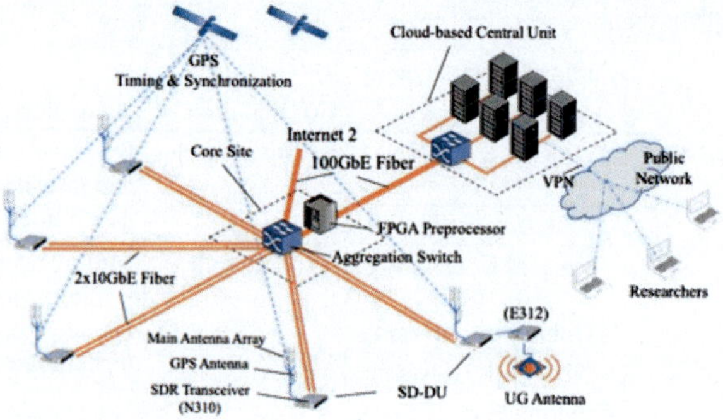

Fig. 4.1 CoSeC-RAN architecture

communication, and other real-life configuration for mobile devices. Moreover, it also supports coordinated multipoint (CoMP), Distributed spectrum sensing, and Cooperative MIMO (CO-MIMO). Finally, it can also be configured using existing protocols and standard to interface with off-the-shelf wireless devices. This ability allows for experiments at the application level [38].

4.3.1.3 Site Selection

The site selection for 5 SD-DUs is not random but strategic. They are selected with goal to support diverse wireless technologies. It tries to capture urban environment, vehicular, underground, and open environment. To that end, three sites were selected on two campuses of Nebraska Lincoln to test for urban environment, and one street site tests the vehicular-to-infrastructure (V2I) wireless experiments and underground to above-ground (UG-AG) wireless communication via MIMO antenna and SDR transceiver. Finally, to test for open environment, one site is selected on rooftop of Nebraska Innovation Campus. mmWave and Terahertz communication are reserved to improve the architecture for future purposes [39, 40].

4.3.1.4 Ethernet and Cloud Network

A fronthaul Ethernet network connects SD-DU with central unit. There are two lines of 10 Gbps for this purpose. Moreover, high-end FPGA cards are used to provide interfacing between Ethernet network and a different base station using eCPRI protocol. The throughput of these FPGAs is 400 Gbps. With such high throughput, direct experiments on Ethernet network can also be conducted.

The cloud central unit at HCC consists of many CPUs and GPUs providing high computational resources for researchers. HCC already has an experience with several RFML projects [79, 80]. Since there is a direct connection between CPUs/GPUs and SD-DU, online experiments of RFML can also be conducted.

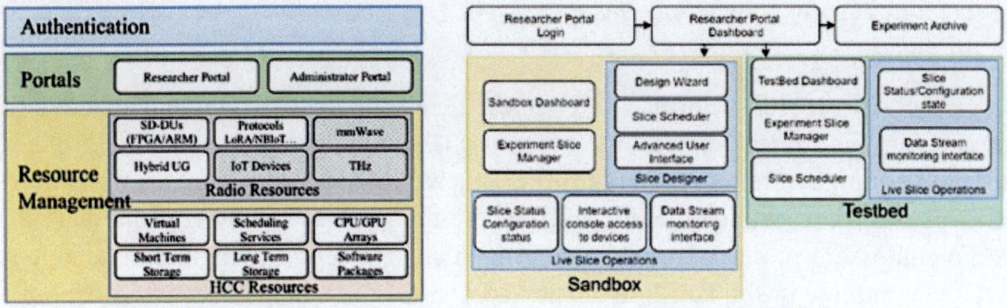

(a) Software System Architecture (b) Research User Workflow

Fig. 4.2 CoSeC-RAN workflow

4.3.2 Software Architecture

The software architecture of CoSeC-RAN uses two types of resources: Radio and HCC resources. The software architecture will be completed in stages where radio resources will be developed new completely from the scratch. HCC resources, however, will be developed using existing components. These components will be customized as per the requirement of CoSeC-Ran. The end result will be a scalable and reusable stateless system [41–45].

The software architecture of CoSeC-RAN is a layered architecture shown in Fig. 4.2a. The first layer provides authentication, the second layer provides front-end, and the third layer performs resource management for all the resources mentioned in Sect. 4.3.1. The front-end is a web portal which gives different aspects of CoSeC-RAN. The purpose of a researcher portal is to create an abstraction layer between the user and the system and hide the necessary details of the system. The user only needs to know the objective and experiment goal to be able to use the system. The administrator portal allows the user to control and administer the system.

The resource management layer has all the resources that will be used by the system. This layer lists out some of the very important resources from HCC, e.g., virtual machines, high performance short- and long-term storage, scheduling services, and CPU/GPU arrays. Virtual machines can be configured as servers and storage are used for archiving data and live stats of the experiment. Moreover, scheduling services are used to manage allocation of CPU time for resources on all levels. Similarly, radio resources are also shown in Fig. 4.2a. These radio resources are expected to support wireless use cases of CoSeC-RAN.

Several tutorials to reserve CoSeC-RAN resources will be provided to researchers so that they can reserve and configure the system as per their need. System will then run the experiments as per scheduled time after configuring the testbed with reserved resources.

4.3.3 Researcher Workflow

CoSeC-RAN provides four basic functionalities for their tests: staging, scheduling, monitoring, and retrieving results. Staging phase includes running short scripts to test and evaluate the simulation platform. This will then be translated to experiments in the sandbox environment (Fig. 4.2b) where the user will be able to interact with radio equipment in that particular experiment slice. Users will login and be able to see the complete dashboard to manage and control the experiments with live high-level stats [46–52].

There is a slice designer module in the system which helps the user to set up and allocate the resources for an experiment. For this purpose, a Design Wizard, CoSeC-RAN Profiles, optimizes experiment for any specific use case. The profiles will have the ability to store network topologies, images of disk or radio, and data files. All of this will be saved as a single file and the researcher will upload the file to get the configurations. Moreover, it will also have the ability to load individual FPGA images and device files. This design wizard will be improved and keep updating throughout the lifetime of CoSeC-RAN. Moreover, the user will be able to change the components that were automatically generated from the design wizard, and even they can create a slice from scratch. However, building slice from scratch is only limited to advanced users. Once the experiment is defined, slice is reserved in a sandbox, and its slice can be deployed in live testbed.

CoSeC-RAN also provides the functionality of reproducing the experiments, both online and offline. Online reproduction involves the user reserving the same resources and loading their profiles from administration portal. The loaded profiles combined with the input data files can then be used to reproduce experiments from the past. Offline reproduction only reproduces the computational process from the research software and already published data files. This reproduction capability helps to test and compare the new ideas with old ones. HCC provides the storage for large archival datasets (in several petabytes). This dataset will be kept best effort but not backed up. This open radio dataset can further simplify the development of RFML. Users will have choice to use either standard or customized data storage plans.

4.4 Testbed Implementation

The section will discuss the technical details of the testbed.

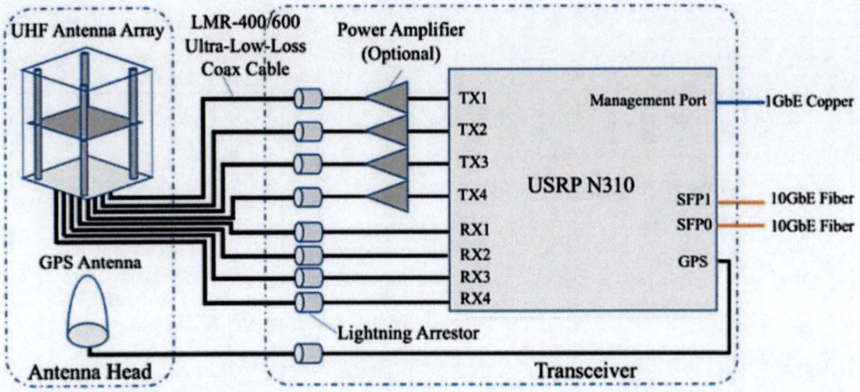

Fig. 4.3 SD-DU schematic

4.4.1 Software-Defined Distribution Unit (SD-DU)

SD-DU is a software-defined transceiver which performs the conversion between digital baseband IQ data and radio waves. Figure 4.3 shows the schematic diagram of SD-DU. SD-DU uses a Universal Software Radio Peripheral (USRP) N310 [74] as a software-defined radio (SDR) transceiver. SDR has four transmission and four receiving channels. Each channel has a bandwidth of 100 megahertz. It has a frequency of 10-6 megahertz and a maximum power of transmission line (Tx) between 12–18 decibel milli.

Each of the transmission and receiving channels in SDR connects with the transmitting and receiving elements of UHF antenna array. The connection is made using ultra-low loss coaxial cables of LMR-400/600 grade. This cable has, for a 50 ft length, insertion loss of 1.2/0.6 dB and 3.2/1.6 dB at 600 megahertz and 2.4 gigahertz, respectively. The cable length may differ from site to site, and however, it remains between 25 and 50 ft. A 3.3 V GPS antenna provides clock synchronization with an SDR receiver. A lightening arrestor is also added to protect the outdoor UHF antenna and GPS antenna. The arrestor is added with RF and GPS ports [53].

A 2 $\times$ 2 uniform rectangular UHF antenna array is designed to cover sub-1 GHz band with an omnidirectional monopole antenna element and 6 dBi gain. This antenna element is an off-the-shelf component. The antenna array can do beamforming between 428 and 856 megahertz. To provide the proof of concept for underground wireless communication, the street site has an additional SD-DU, USRP E312, with a UG antenna. USRP is a 2 $\times$ 2 MIMO SDR with 1 Gbps Ethernet port. UG antenna is shielded to make it waterproof and buried underground at the depth of 30 centimeter.

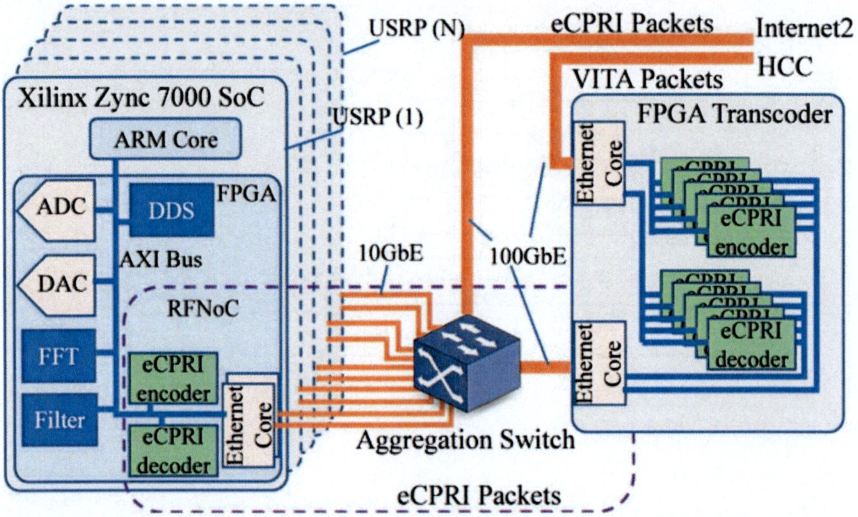

Fig. 4.4 Fronthaul network schematic

4.4.2 *Fronthaul Network*

A dedicated high-speed Ethernet network, over optical fiber, is used to provide secure high-speed Internet and low latency. Equipment used with the network includes an aggregation switch, FPGA preprocessor, and dedicated fiber. Each aggregation switch connects with the SD-DU, HCC, and Internet 2 [7]. Internet 2 connects CoSeC-RAN with 210+ U.S. educational institutes, corporations, and government agencies and their testbeds. Mellanox SN2410 [15] is used as an aggregation switch. The switch has 48 downlink and 8 uplink ports and a throughput of 100 Gbps and can perform Layer 2 and Layer 3 forwarding. FPGA host server has a 3.1 GHz Intel Celeron Dual Core processor with 8 GB DDR4 RAM, 250 GB SSD hard drive, and 850 Watt power module. Both, switch and FPGA server, are in Walter Scott Engineering Center (WSEC). The FPGA program in USRP is based on Radio Frequency Network on Chip (RFNoC) [12]. The RFNoC architecture connects different modules via AXI bus as shown in the left side of Fig. 4.4. Moreover, it is possible to add custom modules to FPGA image of USRP so that it becomes compatible with Cloud-Radio Access Network (C-RAN). To that end, custom modules should be able to translate VRT protocol to Common Public Radio Interface over Ethernet (eCPRI) [4]. The FPGA preprocessor can do backward translation from eCPRI to VITA [54–60, 75].

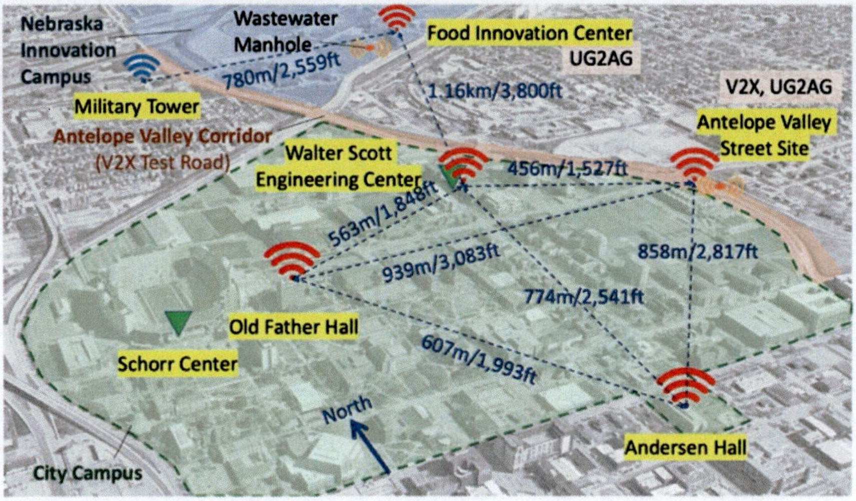

Fig. 4.5 Radio (red) and data site (green) and future sites for expansion (blue)

4.4.3 *Cloud-Based Central Unit*

HCC [8] provides many high performance computing (HPC) resources like dedicated virtual machines (VMs), SLURM-based HPC cluster [71], Open Science Grid (OSG) [16] and distributed data file systems for preprocessing and data storage. Researchers have to use a virtual private network to connect with the HCC facility. Direct layer 2 connection with SD-DU can be established using VPN. However, researcher do not need VPN, if HCC resources are being used directly on testbed.

HCC clusters can be used for applications requiring online baseband processing and/or high radio bandwidth. Researcher can run tens of thousands of jobs in parallel for online baseband processing and OSG, to run jobs, for offline processing. Rhino cluster, located with aggregation switch, can be used for this purpose. This cluster has latency of less than 1 milli second between SD-DU and computing nodes. For RFML, Crane cluster can be used to access air-interfaces of SD-DU. Data from experiments can be stored on HCC for future usage. CoSeC-RAN will become a data repository of RF where researcher will be able to publish data from their own experiments and/or can retrieve results/data published from past experiments for comparison or reproduction [61–66].

4.4.4 *Site Planning and Deployment*

There is a total of seven sites used (five radio and two data). Figure 4.5 shows the location of each site and line-of-sight (LoS) distance between these sites.

Four radio sites are established on rooftop of different buildings of University of Nebraska Lincoln and one radio site is established as a street site at the intersection of Antelope Valley Parkway and Vine street. The distance between radio sites is calculated in a way to avoid correlation of results from spectrum sensing experiments. However, the LoS distance is kept under 1 km to establish a successful wireless connection between sites with a link budget of 600 megahertz. The distance of 1 km is chosen to support mmWave researches in future (Fig. 4.6).

For the campus site, SDR transceivers are kept indoor and antenna heads are mounted at the pole on rooftop, whereas for street side, transceivers are secured in weatherproof shield and antenna heads are mounted at the traffic signal pole. Power is supplied through the traffic signal cabinet and a UG antenna is deployed 5 ft underground next to the cabinet. All five SD-DUs are deployed in a way that they are located within the area of some cellular base stations with a variety of wireless channels, e.g., AH and WSEC and FIC and WSEC, etc., existing between radios sites. The LoS distance between sites ranges from 450 to 940 meter. Such planning support CoSeC-RAN future expansion for 5G experiments, point-to-point sub-6 GHz and mmWave communication. On the street site, the LoS distance is 12 m for underground to above-ground (UG2AG) communication.

4.5 Scalability

CoSeC-RAN has the ability to scale both in frequency and in spatial domain. More sub-6 GHz bands can be accommodated by replacing the antenna head of SD-DU with different antenna bands. Moreover, to accommodate higher frequencies, more broadband amplifiers can be used. The future plan involves deploying radios on military sites to test for mmWave communication and long-distance THz wireless communication.

As far as the spatial domain is concerned, the current fronthaul network of 200 Gbps has the capacity of supporting at-least five more SD-DUs. To add more, it is required to install more switches. The current fiber network between HCC and core site can support 60 SD-DUs without new lines being installed. This fiber has the capacity of 1.2 tera bit per second. The street site has three extra pair of fiber lines with each pair supporting a maximum of 10 sites.

Federal Communication Commission has given access to experimental spectrum zone for this testbed. No license is needed for low-power experiments. For experiments with high transmit power, UNL has access to experimental spectrum WA3XCD.

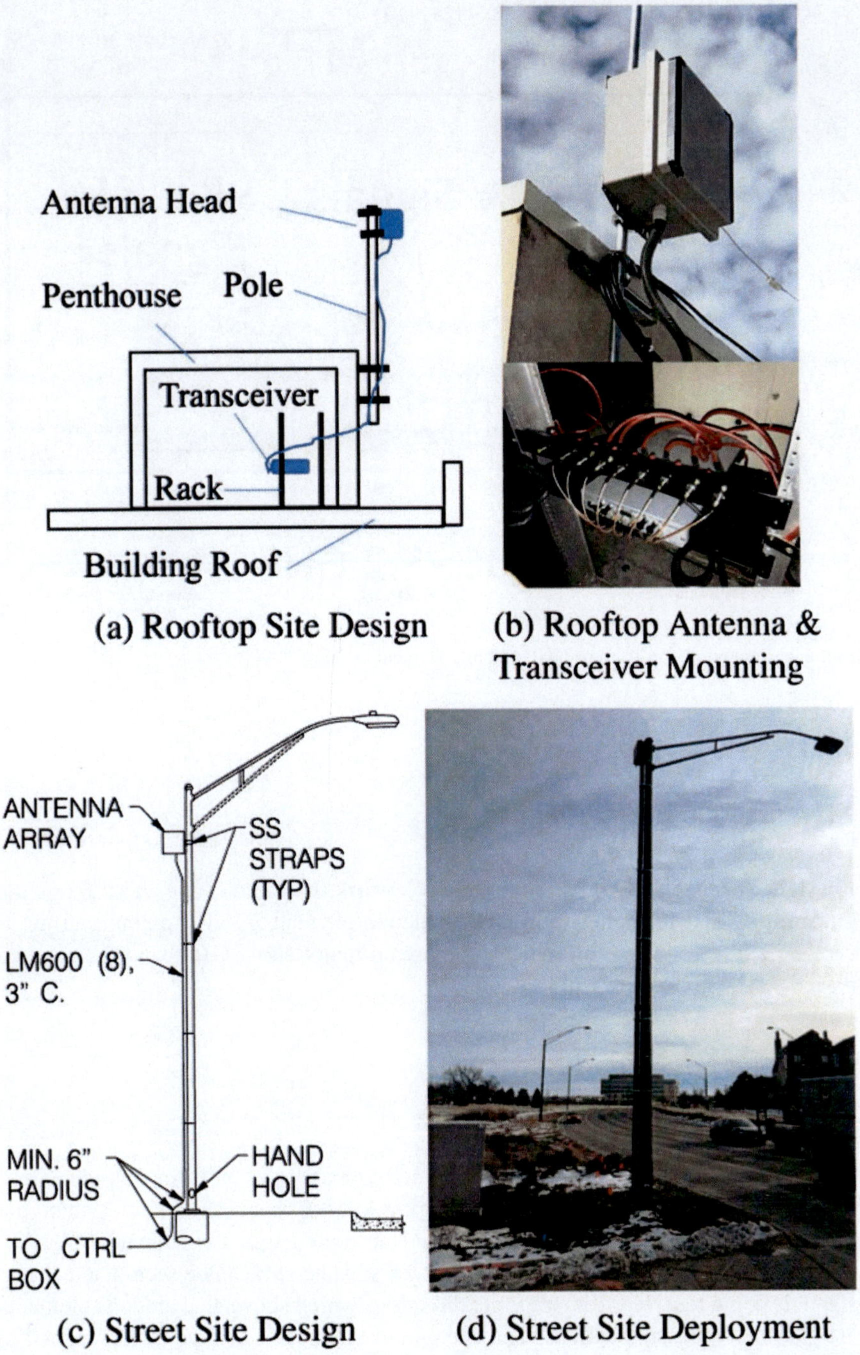

(a) Rooftop Site Design (b) Rooftop Antenna & Transceiver Mounting

(c) Street Site Design (d) Street Site Deployment

Fig. 4.6 Campus and street site implementation

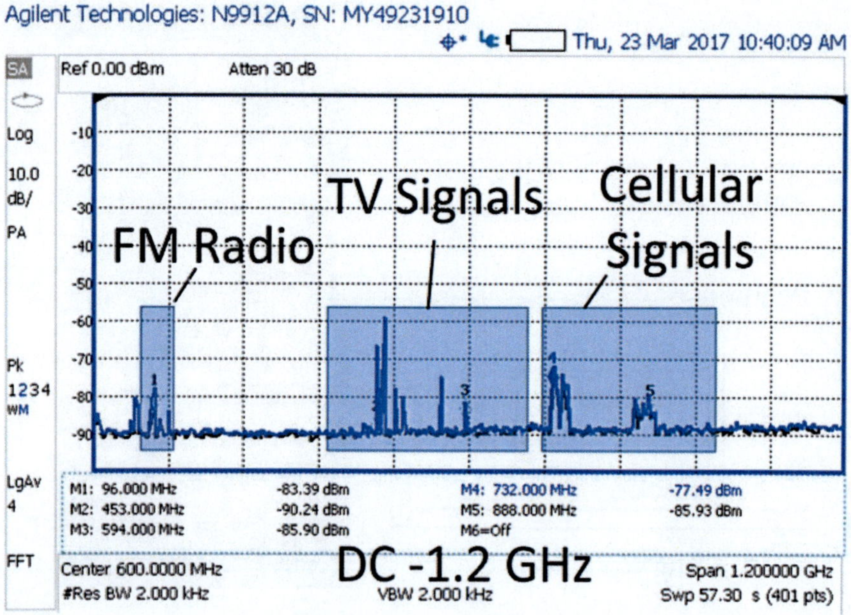

Fig. 4.7 Receiving spectrum of UHF antenna element in DC-1.2 GHz

4.6 Testing and Demonstration

4.6.1 Antenna Performance

Keysight Spectrum (N9912A) and vector Network (N9923A) are used to gauge the transmitting and receiving prowess of antenna element in UHF antenna array. It can be seen from Fig. 4.7 that it can capture all major sub-1 GHz bands even when placed on the ground [67].

4.6.2 Receive Beamforming

To test receiver beamforming, the digital baseband signal from antenna array is collected and processed offline using MATLAB beamformer, and then the beamformed signal is plotted in H-plane. The plotted signal is shown in Fig. 4.8. The results are from **6** MHz and four local TV channels. It can be seen that a strong signal is being received from multiple directions which shows that antenna elements are placed at optimal locations. The strength of the received signal ranges from **4** to 6 dB showing that UHF array has a very good directivity.

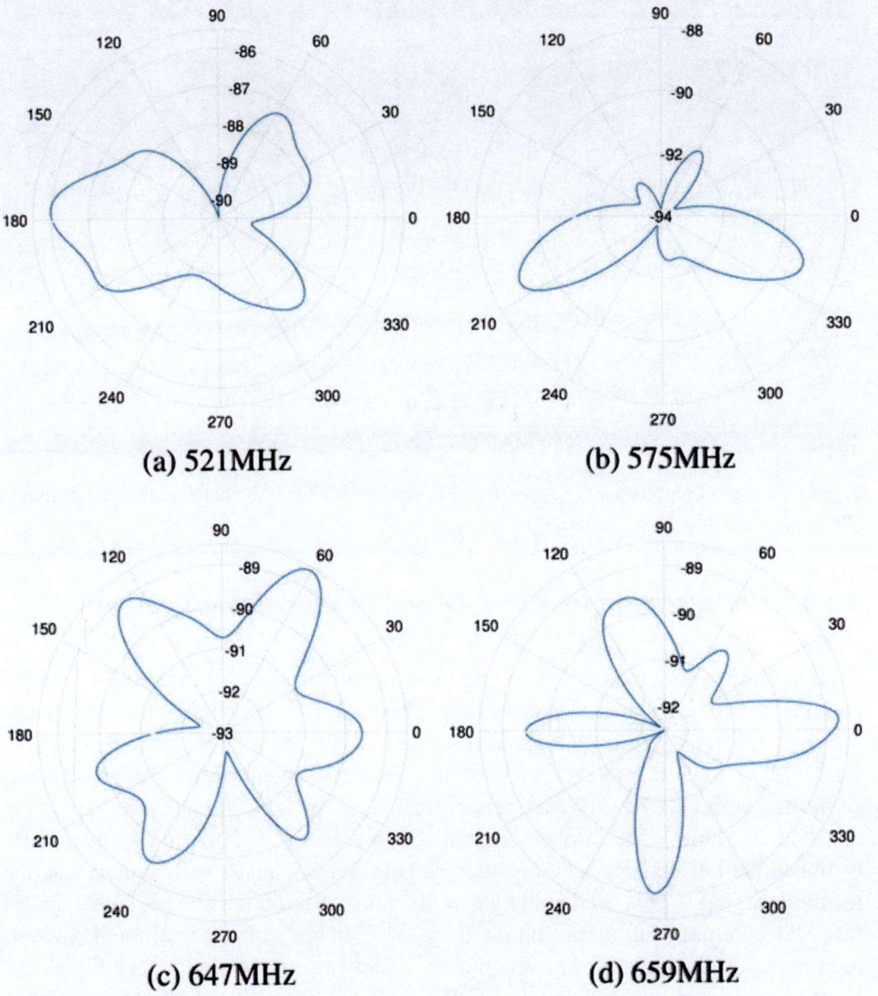

Fig. 4.8 Receive beamforming on 2 × 2 uniform rectangular array, H-plane scan of 4 local TV channels with 6 MHz bandwidth, polar axis unit: dBm

4.6.3 Distributed Spectrum Sensing

GNU radio receives data from all five sites at the same time and plots the spectra-temporal pattern via waterfall plot. Figure 4.9 shows the spectrum sensing at every 4 s. The results are shown as a waterfall plot (above) and instantaneous spectrum (below). It can be seen that multiple strong narrow-band signals are received at all five sites. The strength of the signals is between **433** and 433.6 MHz [69].

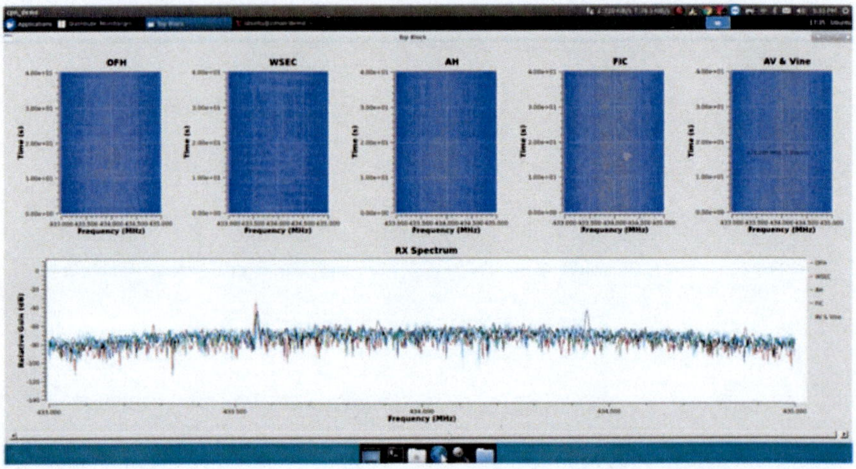

Fig. 4.9 Waterfall plot (above) and spectrum (below) of 433–435 MHz band on five campus radio sites

4.6.4 Underground to Above-Ground Communication

Underground communication is being used in infrastructure monitoring [1], agriculture [3, 78], and border patrol [72]. The UG2AG experiment is carried out at the street site with a UG antenna buried at 30 centimeter and connected to USRP E312 SDR transceiver. AG antenna is kept 7 meter above the ground and connected with USRP N310 transceiver. Both, UG and AG, antennas are kept at the distance of 10 m. A 500 kilohertz signal with a power of 5 dBm and 43 megahertz is transmitted from UG to AG antenna (Fig. 4.6). Due to soil attenuation the receiving signal is only 30 decibel above the noise floor. Such a weak signal shows that UG communication is difficult from over-the-air communication. However, modifications like high transmitting power and advanced modulation techniques can be used to increase the performance [70].

4.6.5 Vehicle-to-Infrastructure (V2I) Communication

The V2I test is performed on a Sedan car on the street site. Two antennas, 5.8 gigahertz and UHF antenna, are used for the performance test. Both antennas were mounted outside the car as shown in Fig. 4.10. USRP B200 is used inside the

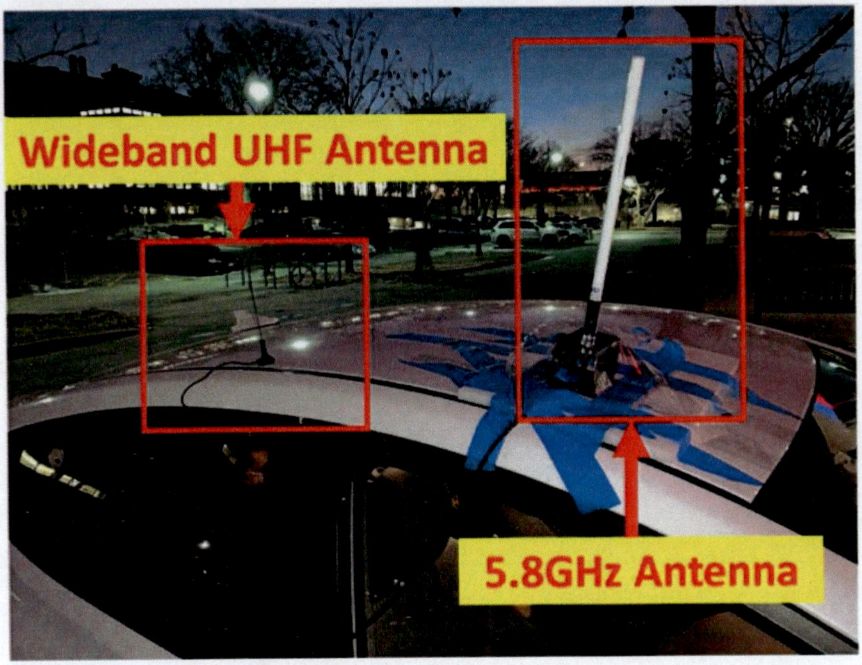

Fig. 4.10 V2I communication test setup for vehicular antennas

car. The experiment used OFDM signal of **30** dBm and 500 kilohertz bandwidth. Two experiments were performed with two different central frequencies of 915 megahertz and 5.8 gigahertz. Bit error rate (BER) and RSS are plotted in Figs. 4.11a and b for 90s. BER gaps represent the dropped frames due to non-synchronization. The average BER for 5.8 gigahertz is 0.0079 and that for 915 megahertz is 0.0262. BER is better for 5.8 gigahertz because of higher noise floor. RSS is 50 decibel higher for 915 megahertz as compared to 5.8 gigahertz. RSS is dependent upon the distance between both transmitter and receiver.

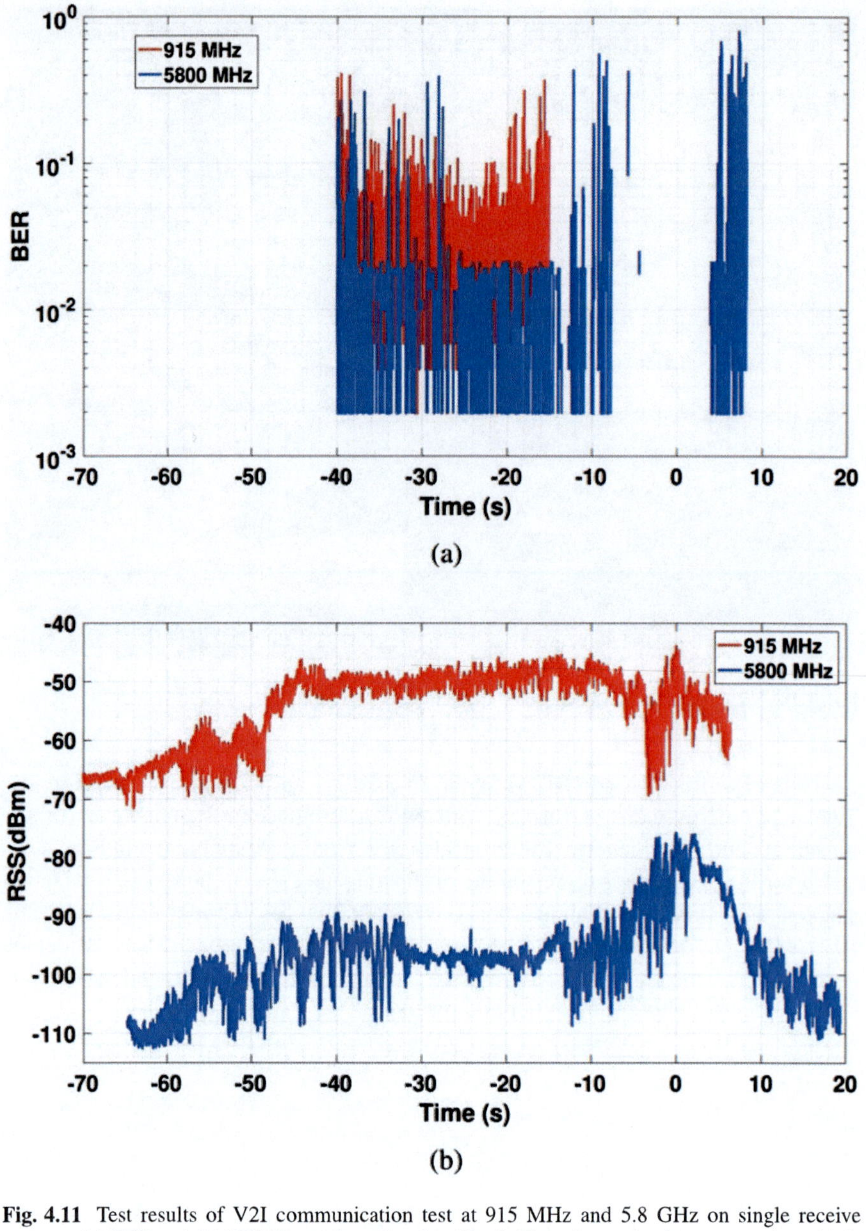

Fig. 4.11 Test results of V2I communication test at 915 MHz and 5.8 GHz on single receive antenna: (**a**) BER and (**b**) RSS. Route from South to North. Time 0 s marks the moment the vehicle passing by the road light pole of the receiver antenna

References

1. I.F. Akyildiz, D. Pompili, T. Melodia, Underwater acoustic sensor networks: research challenges. Ad Hoc Netw. **3**(3), 257–279 (2005)
2. J. Breen, A. Buffmire, J. Duerig, K. Dutt, E. Eide, A. Ghosh, M. Hibler, D. Johnson, S.K. Kasera, E. Lewis, et al., Powder: Platform for open wireless data-driven experimental research. Comput. Netw. **197**, 108281 (2021)
3. X. Dong, M.C. Vuran, S. Irmak, Autonomous precision agriculture through integration of wireless underground sensor networks with center pivot irrigation systems. Ad Hoc Netw. **11**(7), 1975–1987 (2013)
4. A.B. Ericsson et al., Common public radio interface (CPRI); interface specification v7. 0. *Huawei Technologies Co. Ltd, NEC Corporation, Alcatel Lucent, and Nokia Networks* (2015)
5. M. Ettus, M. Braun, The universal software radio peripheral (USRP) family of low-cost SDRs. *Opportunistic Spectrum Sharing and White Space Access: The Practical Reality*, pp. 3–23 (2015)
6. S. Gokceli, G.K. Kurt, E. Anarim, Cognitive radio testbeds: State of the art and an implementation. *Spectrum Access and Management for Cognitive Radio Networks*, pp. 183–210 (2017)
7. Home. https://internet2.edu, Jan 2022. [Online; accessed 27. Jan. 2022]
8. Holland Computing Center | Nebraska. https://hcc.unl.edu, Jan 2022. [Online; accessed 27. Jan. 2022]
9. A. Konda, A. Rau, M.A. Stoller, J.M. Taylor, A. Salam, G.A. Pribil, C. Argyropoulos, S.A. Morin, Soft microreactors for the deposition of conductive metallic traces on planar, embossed, and curved surfaces. Adv. Funct. Mater. **28**(40), 1803020 (2018)
10. J.-P. Lang, Gnu radio-the free and open software radio ecosystem. GNU Radi0 (2013)
11. A. Loch, M. Schulz, M. Hollick, Warp drive-accelerating wireless multi-hop cross-layer experimentation on SDRs, in *Proceedings of the 2014 ACM Workshop on Software Radio Implementation Forum*, pp. 31–36 (2014)
12. J. Malsbury, M. Ettus, Simplifying FPGA design with a novel network-on-chip architecture, in *Proceedings of the Second Workshop on Software Radio Implementation Forum*, pp. 45–52 (2013)
13. V. Marojevic, D. Chheda, R.M. Rao, R. Nealy, J.-M. Park, J.H. Reed, Software-defined LTE evolution testbed enabling rapid prototyping and controlled experimentation, in *2017 IEEE Wireless Communications and Networking Conference (WCNC)* (IEEE, 2017), pp. 1–6
14. T.R. Newman, A. He, J. Gaeddert, B. Hilburn, T. Bose, J.H. Reed, Virginia tech cognitive radio network testbed and open source cognitive radio framework, in *2009 5th International Conference on Testbeds and Research Infrastructures for the Development of Networks & Communities and Workshops* (IEEE, 2009), pp. 1–3
15. NVIDIA Spectrum SN2000 Open Ethernet Switches. https://www.nvidia.com/en-us/networking/ethernet-switching/spectrum-sn2000, Jan 2022. [Online; accessed 27. Jan. 2022]
16. OSG. https://opensciencegrid.org, Jan 2022. [Online; accessed 27. Jan. 2022]
17. PAWR NSF, Cosmos: Cloud enhanced open software defined mobile wireless testbed for city-scale deployment (2023)
18. Platforms for advanced wireless research (Dec 2021)
19. D. Raychaudhuri, I. Seskar, M. Ott, S. Ganu, K. Ramachandran, H. Kremo, R. Siracusa, H. Liu, M. Singh, Overview of the orbit radio grid testbed for evaluation of next-generation wireless network protocols, in *IEEE Wireless Communications and Networking Conference, 2005*, vol. 3 (IEEE, 2005), pp. 1664–1669
20. U. Raza, A. Salam, On-site and external power transfer and energy harvesting in underground wireless. Electronics **9**(4), 681 (2020)
21. U. Raza, A. Salam, A survey on subsurface signal propagation. Smart Cities **3**(4), 1513–1561 (2020)

22. U. Raza, A. Salam, Wireless underground communications in sewer and stormwater overflow monitoring: Radio waves through soil and asphalt medium. Information **11**(2), 98 (2020)
23. U. Raza, A. Salam, Zenneck waves in decision agriculture: An empirical verification and application in EM-based underground wireless power transfer. Smart Cities **3**(2), 308–340 (2020)
24. A. Salam, Pulses in the sand: Long range and high data rate communication techniques for next generation wireless underground networks. *ETD Collection for University of Nebraska - Lincoln* (AAI10826112, 2018)
25. A. Salam, A comparison of path loss variations in soil using planar and dipole antennas, in *2019 IEEE International Symposium on Antennas and Propagation* (IEEE, Jul 2019)
26. A. Salam, Design of subsurface phased array antennas for digital agriculture applications, in *Proc. 2019 IEEE International Symposium on Phased Array Systems and Technology (IEEE Array 2019)*, Waltham, MA, USA, October 2019
27. A. Salam, A path loss model for through the soil wireless communications in digital agriculture, in *2019 IEEE International Symposium on Antennas and Propagation* (IEEE, Jul 2019)
28. A. Salam, Sensor-free underground soil sensing, in *ASA, CSSA and SSSA International Annual Meetings (2019)* (ASA-CSSA-SSSA, 2019)
29. A. Salam, Subsurface MIMO: A beamforming design in internet of underground things for digital agriculture applications. J. Sensor Actuat. Netw. **8**(3), 41 (2019)
30. A. Salam, *Underground Environment Aware MIMO Design Using Transmit and Receive Beamforming in Internet of Underground Things* (Springer International Publishing, Cham, 2019), pp. 1–15
31. A. Salam, An underground radio wave propagation prediction model for digital agriculture. Information **10**(4), 147 (2019)
32. A. Salam, Underground soil sensing using subsurface radio wave propagation, in *5th Global Workshop on Proximal Soil Sensing*, COLUMBIA, MO, May 2019
33. A. Salam, *Internet of Things for Environmental Sustainability and Climate Change* (Springer International Publishing, Cham, 2020), pp. 33–69
34. A. Salam, *Internet of Things for Sustainability: Perspectives in Privacy, Cybersecurity, and Future Trends* (Springer International Publishing, Cham, 2020, pp. 299–327
35. A. Salam, *Internet of Things for Sustainable Community Development*, 1st edn. (Springer Nature, New York, 2020)
36. A. Salam, *Internet of Things for Sustainable Community Development: Introduction and Overview* (Springer International Publishing, Cham, 2020), pp. 1–31
37. A. Salam, *Internet of Things for Sustainable Forestry* (Springer International Publishing, Cham, 2020), pp. 147–181
38. A. Salam, *Internet of Things for Sustainable Human Health* (Springer International Publishing, Cham, 2020), pp. 217–242
39. A. Salam, *Internet of Things for Sustainable Mining* (Springer International Publishing, Cham, 2020), pp. 243–271
40. A. Salam, *Internet of Things for Water Sustainability* (Springer International Publishing, Cham, 2020), pp. 113–145
41. A. Salam, *Internet of Things in Agricultural Innovation and Security* (Springer International Publishing, Cham, 2020), pp. 71–112
42. A. Salam, *Internet of Things in Sustainable Energy Systems* (Springer International Publishing, Cham, 2020), pp. 183–216
43. A. Salam, *Internet of Things in Water Management and Treatment* (Springer International Publishing, Cham, 2020), pp. 273–298
44. A. Salam, A.D. Hoang, A. Meghna, D.R. Martin, G. Guzman, Y.H. Yoon, J. Carlson, J. Kramer, K. Yansi, M. Kelly, et al., The future of emerging IoT paradigms: Architectures and technologies (2019)
45. A. Salam, U. Karabiyik, A cooperative overlay approach at the physical layer of cognitive radio for digital agriculture, in *Third International Balkan Conference on Communications*

and Networking 2019 (BalkanCom'19), Skopje, Macedonia, the former Yugoslav Republic of, June 2019
46. A. Salam, U. Raza, *Autonomous Irrigation Management in Decision Agriculture* (Springer International Publishing, Cham, 2020), pp. 379–398
47. A. Salam, U. Raza, *Current Advances in Internet of Underground Things* (Springer International Publishing, Cham, 2020), pp. 321–356
48. A. Salam, U. Raza, *Decision Agriculture* (Springer International Publishing, Cham, 2020), pp. 357–378
49. A. Salam, U. Raza, *Electromagnetic Characteristics of the Soil* (Springer International Publishing, Cham, 2020), pp. 39–59
50. A. Salam, U. Raza, *Modulation Schemes and Connectivity in Wireless Underground Channel* (Springer International Publishing, Cham, 2020), pp. 125–166
51. A. Salam, U. Raza, On burial depth of underground antenna in soil horizons for decision agriculture, in *2020 International Conference on Internet of Things (ICIOT-2020)*, Honolulu, USA, September 2020
52. A. Salam, U. Raza, *Signals in the Soil*, 1st edn. (Springer Nature, New York, 2020)
53. A. Salam, U. Raza, *Signals in the Soil: An Introduction to Wireless Underground Communications* (Springer International Publishing, Cham, 2020), pp. 3–38
54. A. Salam, U. Raza, *Signals in the Soil: Subsurface Sensing* (Springer International Publishing, Cham, 2020), pp. 251–297
55. A. Salam, U. Raza, *Signals in the Soil: Underground Antennas* (Springer International Publishing, Cham, 2020), pp. 189–215
56. A. Salam, U. Raza, *Soil Moisture and Permittivity Estimation* (Springer International Publishing, Cham, 2020), pp. 299–317
57. A. Salam, U. Raza, *Underground Phased Arrays and Beamforming Applications* (Springer International Publishing, Cham, 2020), pp. 217–248
58. A. Salam, U. Raza, *Underground Wireless Channel Bandwidth and Capacity* (Springer International Publishing, Cham, 2020), pp. 167–188
59. A. Salam, U. Raza, *Variable Rate Applications in Decision Agriculture* (Springer International Publishing, Cham, 2020), pp. 399–423
60. A. Salam, U. Raza, *Wireless Underground Channel Modeling* (Springer International Publishing, Cham, 2020), pp. 61–121
61. A. Salam, S. Shah, Internet of things in smart agriculture: Enabling technologies, in *2019 IEEE 5th World Forum on Internet of Things (WF-IoT) (WF-IoT 2019)*, Limerick, Ireland, April 2019
62. A. Salam, M.C. Vuran, Impacts of soil type and moisture on the capacity of multi-carrier modulation in internet of underground things, in *Proc. of the 25th ICCCN 2016*, Waikoloa, Hawaii, USA, Aug 2016
63. A. Salam, M.C. Vuran, EM-based wireless underground sensor networks, in *Underground Sensing*, ed. by S. Pamukcu, L. Cheng, (Academic Press, Cambridge, 2018), pp. 247–285
64. A. Salam, M.C. Vuran, Smart underground antenna arrays: A soil moisture adaptive beamforming approach, in *Proc. IEEE INFOCOM 2017*, Atlanta, USA, May 2017
65. A. Salam, M.C. Vuran, Wireless underground channel diversity reception with multiple antennas for internet of underground things, in *Proc. IEEE ICC 2017*, Paris, France, May 2017
66. A. Salam, M.C. Vuran, X. Dong, C. Argyropoulos, S. Irmak, A theoretical model of underground dipole antennas for communications in internet of underground things. IEEE Trans. Antenn. Propagat. **67**(6), 3996–4009 (2019)
67. A. Salam, M.C. Vuran, S. Irmak, Pulses in the sand: Impulse response analysis of wireless underground channel, in *The 35th Annual IEEE International Conference on Computer Communications (INFOCOM 2016)*, San Francisco, USA, April 2016
68. A. Salam, M.C. Vuran, S. Irmak, Towards internet of underground things in smart lighting: A statistical model of wireless underground channel, in *Proc. 14th IEEE International Conference on Networking, Sensing and Control (IEEE ICNSC)*, Calabria, Italy, May 2017

69. A. Salam, M.C. Vuran, S. Irmak, Di-sense: In situ real-time permittivity estimation and soil moisture sensing using wireless underground communications. Comput. Netw. **151**, 31–41 (2019)
70. A. Salam, M.C. Vuran, S. Irmak, A statistical impulse response model based on empirical characterization of wireless underground channel. IEEE Trans. Wirel. Commun. **19**, 5966–5981 (2020)
71. Slurm Workload Manager - Documentation. https://slurm.schedmd.com, Dec 2021. [Online; accessed 27. Jan. 2022]
72. Z. Sun, P. Wang, M.C. Vuran, M.A. Al-Rodhaan, A.M. Al-Dhelaan, I.F. Akyildiz, BorderSense: Border patrol through advanced wireless sensor networks. Ad Hoc Netw. **9**(3), 468–477 (2011)
73. S. Temel, M.C. Vuran, M.M.R. Lunar, Z. Zhao, A. Salam, R.K. Faller, C. Stolle, Vehicle-to-barrier communication during real-world vehicle crash tests. Comput. Commun. **127**, 172–186 (2018)
74. USRP N310, An National Instruments Brand. https://www.ettus.com/all-products/usrp-n310, Jan 2022. [Online; accessed 27. Jan. 2022]
75. VITA - Home. https://www.vita.com, Jan 2022. [Online; accessed 27. Jan. 2022]
76. M.C. Vuran, A. Salam, R. Wong, S. Irmak, Internet of underground things in precision agriculture: Architecture and technology aspects. Ad Hoc Netw. **81**, 160–173 (2018)
77. M.C. Vuran, A. Salam, R. Wong, S. Irmak, Internet of underground things: Sensing and communications on the field for precision agriculture, in *2018 IEEE 4th World Forum on Internet of Things (WF-IoT) (WF-IoT 2018)*, Singapore, February 2018
78. M.C. Vuran, A. Salam, R. Wong, S. Irmak, Internet of underground things in precision agriculture: Architecture and technology aspects. Ad Hoc Netw. **81**, 160–173 (2018)
79. K. Yang, M.C. Vuran, S. Scott, F. Guo, C.R. Ahn, Neural network-based channel estimation for 2x2 and 4x4 MIMO communication in noisy channels, in *International Balkan Conference on Communications and Networking* (2018)
80. Z. Zhao, M.C. Vuran, F. Guo, S. Scott, Deep-waveform: A learned OFDM receiver based on deep complex convolutional networks. e-prints (2018). arXiv–1810
81. Z. Zhao, M.C. Vuran, B. Zhou, M.M.R. Lunar, Z. Aref, D.P. Young, W. Humphrey, S. Goddard, G. Attebury, B. France, A city-wide experimental testbed for the next generation wireless networks. Ad Hoc Netw. **111**, 102305 (2021)

Chapter 5
Stormwater Management Modeling (SWMM)

Check for updates

5.1 Introduction

Old combined sewer systems are used to direct waste- and stormwater directly into the waterways. Over the period of time, they have been extended to send it to Wastewater Treatment Plant (WWTP). However, even today, due to blockages or excess stormwater, wastewater is still directed to rivers and streams. This phenomenon is known as combined sewer overflow (CSO). CSO has adverse effect on health and economy, contaminates water, increases the level of coli bacteria (E) in the water, which results in up to 5500 gastrointestinal illnesses per year [3–6, 63].

As per [65], over 850 billion gallons of CSO per year occurs around 750 cities. These CSOs are very expensive to handle, which is the reason that most cities ignore them. However, in accordance with the Clean Water Act policy, EPA issued a CSO Control Policy that mandated cities to present a Long-Term Control Plan (LTCP) to minimize CSOs [7–21].

5.2 Current Methods

The main goal of the cities was to direct the wastewater or stormwater into the WWTP during peak times of inflow all around the year. Cities used SWMM software for the prediction of incoming of wastewater and stormwater giving them an idea of amount of water that can overflow. Data analyzed from SWMM show that CSOs can be minimized or prevented by either increasing the storage capacity of the system or reducing the amount of flow entering the system. Figure 5.1 shows the effect of different combinations of green infrastructure and increased storage capacity over CSOs [66]. The data for this graph are collected from analysis of SWMM data. Figure 5.2 and many other like this help cities to fight with CSOs. For example, from this figure, it can be concluded that 1.6 million gallons of storage

A. Salam, *Internet of Things in Smart Sewer and Drainage Systems*,
https://doi.org/10.1007/978-3-031-48109-3_5

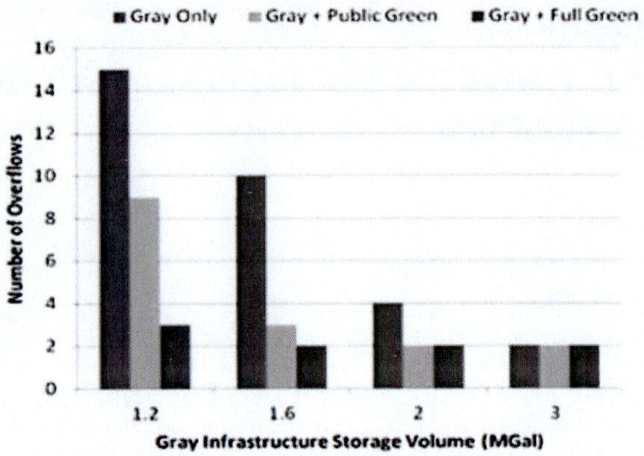

Fig. 5.1 SWMM data of CSOs with feasibility (EPA, 2014b)

capacity along with the robust green infrastructure is more useful and effective than just adding 3 million gallons of storage. Next, we study the work being done using three traditional methodologies, i.e., green infrastructure, Gray infrastructure, and separation, to reduce CSOs.

Separation can completely eliminate CSOs. Grand Rapids reduced their expense of treating wastewater from 2.7 million USD to 1.9 million USD by converting combined sewers into separate sewers over the period of 25 years, i.e., 1991 to 2015 [2]. Moreover, by the year of 2015, they completely eliminated their CSOs from 12 billion gallons in 1970s. However, the major hurdle to the adoption of separation method is its enormous cost. Even the Great Rapids separation cost them around 400 million dollars [2].

Green methods are inexpensive and highly recommended solution by EPA to be included in every CSO reduction plan. Draining is done by both existing (roads, parking lots) and new infrastructure (green spaces and rain gardens). Retention is done through green roofs, retention ponds, and planter trenches. Municipalities incentivize businesses using these methods in terms of low fees, tax, and grants [64]. An advantage of green infrastructure is that it adds green spaces and beautifies the city, therefore improving the overall community. Due to low cost of green solution, Philadelphia allocates 70% of its LTCP with 1.7 billion USD to green methods of preventing CSOs [22–30, 59].

Gray Infrastructure method builds more storage spaces on the way to WWTP or increase the capacity of the plant itself. Age old combined sewer systems, with damaged and cracked leaky pipes, have further worsened the CSOs events. The smoothness of the pipes have deteriorated over the period of time. These pipes can be replaced or made smooth by lining with some smooth material to increase the capacity of the sewer system. Moreover, capacity can also be increased by increasing size of pipes or adding retention basins. For example, Tunnel and Reservoir Plan (TARP) project in Chicago is being built with a budget of 3.8 billion USD adding 2.3 billion gallons of capacity to the system, therefore lowering CSO occurrence from 100 to 50 days in a year [4]. Many other cities can also implement this on smaller scale.

5.3 Smart Control

Wireless enabled sensor technology is emerging technology that aims to control CSOs dynamically considering changing situations within the system. The technology aims to direct the water flow from pipes of overwhelming area to the pipes of less overwhelming areas, hence increasing the overall capacity of the system. This technology uses sensors to collect data and strategically placed pumps, gates and valves, and Linux-based gateways to send data to/from database. The sensor nodes are made from mote, i.e., a low-powered small computer, and come with a storage chip, battery pack, a radio, and probes for water level and flow rate [31–45].

Gateway transfers information to control center, and this information is also used by actuators to control valves, gates, and pumps. Gateway works on multiple control algorithms. For example, one algorithm allows valve to open or close based on the values of pipe capacity. Another allows to calculate the capacity of the pipes so that when the pipe with lowest capacity is full, wastewater or stormwater can be directed to other pipes by opening valves. Similarly, one of the algorithm checks for the weather forecast so that timely arrangements can be made [5].

5.4 Case Study of South Bend, Indiana

South Bend, Indiana, suffered from around 3 billion gallons of CSO between 1997 and 2004. Moreover, they also spent a huge sum of money, 87 million USD, in an attempt to improve their CSO infrastructure. South Bend submitted their LTCP plan in 2004 and began their CSOnet Project that later became the part of their LTCP plan due to its success. CSOnet project is a real-time decision support system (DSS) that aims to optimize the sewer operations. CSOnet uses 150 sensors and 17 gateways to send real-time data to central cloud-based center for analysis. CSOnet utilizes the capacity of existing infrastructure to reduce sewer overflow. Phase 1 of LTCP plan took more than 150 million USD and improvement to CSOnet just cost 6 million USD. This shows how beneficial this project was from financial point of view. Public works department uses Supervisory Control and Data Acquisition (SCADA), IBM's Intelligent Operations Center software, and city's geographic information system (GIS) for real-time monitoring of pipes network, pipe levels, storage capacity of the system, levels of retention basins, and any CSOs that might be happening. A dry weather CSO occurs due to pipes blocked from tree roots and debris. These dry weather CSOs can cost city a lot of money in terms of property damages and fine. Data analysis can help in differentiating the expected flow during regular and peak times. Hence, when the flow level exceeds normal level, blockage can be removed. This preventive measure of dealing CSO has helped South Bend to completely eliminate dry CSO [46–53].

A sensor network is used to stop CSO to enter into St. Joseph River using wires at outfalls and diverting this water to a throttle line ending up at WWTP through

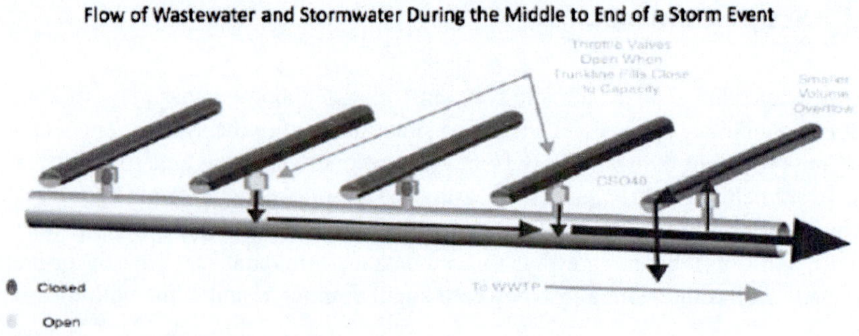

Fig. 5.2 The Competing Algorithm in action

interceptors. It was noticed that some of the interceptor were not operating at their full capacity resulting in overflow from outfalls. Therefore, valves were installed between trunk line and interceptor. These valves can be controlled using a control algorithm. The valves stayed closed until the throttle line was not full and were opened as soon as the capacity of the lines reached threshold value. The process is shown in Fig. 5.2.

Weather prediction in CSOnet allows for preventive measures before storm even happens. If a high stormwater flow is expected, CSOnet dynamically empties the retention basins to river, therefore creating more space for stormwater in the system. CSOnet was implemented in 2008; since then, it reduced the CSOs from 2.1 billion gallons to 458 million gallons by 2014. Moreover, data from SWMM show that CSOnet prevented 25% of the overall CSOs [5].

5.5 Case Study of Richmond, Virginia

Richmond has suffered a lot from combined sewer overflows. On average, Richmond sees a 44 inches of rainfall per year, and 26 CSO outfalls have contributed over 3 billion gallons of CSO in James River [61]. Richmond has invested 242 million USD so far and still has a need to spend another 500 million USD [1]. Moreover, LTCP of Richmond has focused more on extending the storage and treatment capacity by building Gray infrastructure and adding green infrastructure to the ecosystem [62].

Combination of green and Gray infrastructure has been very successful for Richmond. Shockoe Basin, a retention basin, has a capacity of 44 million gallons of water. A couple of interceptors (7-feet diameter) are connected to basin to direct water to WWTP. This resulted in reduced CSO at 10 of the outfalls. Moreover, other attempt of increasing storage capacity includes Hampton–McCloy tunnel, with a capacity of 7 million gallons, and was built with a big budget of 50 million USD. This resulted in reduced CSO on 3 more outfalls near critical locations [54–56].

Table 5.1 A comparison of the costs associated with different control methods based on SWMM

City CSO control	Method	Cost	Drop in CSOs	Cost per million gallon annual drop in CSOs
Richmond, VA	Phase III—gray infrastructure	$375 M	1.44 billion gallons	$260,000
Richmond, VA	Phase III—green infrastructure	$2.6 M	10 million gallons	$260,000
Grand rapids, MI	Sewer improvement project—total separation	$400 M	629 million gallons	$636,000
South bend, IN	CSOnet—wireless sensor technology and throttle line	$6 M	312 million gallons	$19,200

Richmond has made effort to model the stormwater systems by mapping CSO areas and investigating details like material, diameter, and length of the pipe. In an effort to increase green infrastructure, they have created flood zones, added green spaces, and made drainage systems that are part of combined sewer system [11]. Richmond has already completed their Phases I and II of LTCP plan. Phase III plans to improve CSO infrastructure by increasing the capacity of wet weather treatment capacity to 300 million gallons per year, increasing the storage capacity of Shockoe basin by 15 million gallons, extending the storage and chlorine disinfection of troubling outfalls. For green infrastructure, city will improve 18 acres of impervious surfaces including parks, public utility property, city owned properties, and tree boxes [10]. SWMM data show that green infrastructure helped decreasing 10 million gallons CSO annually and Gray infrastructure helped reducing 0.228 to 1.67 billion gallons CSO. A total drop was calculated to be 86% [1]. Table 5.1 shows the cost analysis of CSO after analyzing data from SWMM [57].

5.6 Benefit of Using Wireless Technology in Richmond, VA

The strategies to control combined sewer overflows aim to fight with a lot of adversities such as rainfall, snow melts, blocked pipes, and changing climates, etc. Cities have invested hundreds of millions of dollars in green and Gray infrastructure and are just left with a static solution that is attempting to control dynamic variables involved in sewer overflow, hence overcompensating system to handle peak of SWMM hydrograph. For example, Fig. 5.3 shows the cost versus benefit to analyze the increasing size of interceptor. It was determined that it requires 300 million USD to build a storage capacity that will be fully functional once in every 5 years. Such a big investment for small percentage gain does not make sense. Moreover, it is impossible to achieve 100% control using static systems.

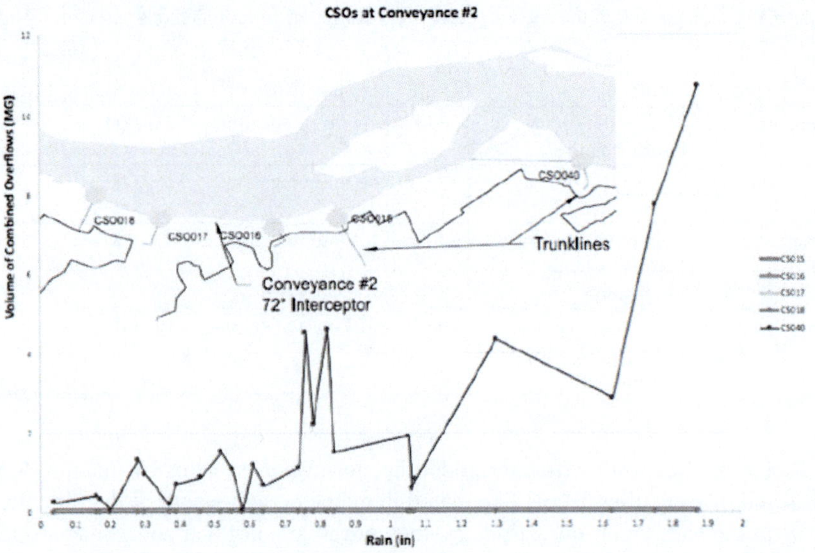

Fig. 5.3 Overflow activity at an interceptor called conveyance number 2, shown as the line that connects the circles representing the 5 outfalls

Richmond has focused more on controlling CSO by monitoring, collecting data, and modeling the data. These efforts can be improved manifold if they add wireless sensors into their LTCP plan. Wireless technology can also be integrated with existing SCADA and GIS capabilities. Moreover, various parameters can be fine-tuned, and components can be modified to make an efficient system. For example, (1) careful data collection can help to identify the crucial pipes that can be replaced, (2) making sure that stormwater enters only through the dam, and (3) controlling flow to improve capacity [58].

CSO's data from SWMM show that Richmond's case is very similar to that of South Bend, i.e., CSOs are occurring on few outfalls connected to interceptor through trunk line. Actuator valves can be used to redirect CSOs to outfalls with small drainage interceptors or, as in the case of south bend, a throttle line can also be added to improve these areas. This can be seen in Fig. 5.3 where after the rainfall of 2 inches, CSO happens at outfall "CSO040" connected to interceptor [59, 60].

References

1. ASCE, 2015 report card for Virginia's infrastructure. Technical report, American Society of Civil Engineers, 2015
2. City of Grand Rapids, Sewer improvement project. Retrieved
3. A. Konda, A. Rau, M.A. Stoller, J.M. Taylor, A. Salam, G.A. Pribil, C. Argyropoulos, S.A. Morin, Soft microreactors for the deposition of conductive metallic traces on planar, embossed, and curved surfaces. Adv. Funct. Mater. **28**(40), 1803020 (2018)

4. M.V. Landis, TARP status report. Technical report, Metropolitan Water Reclamation District of Greater Chicago, 2017
5. L. Montestruque, M.D. Lemmon, Globally coordinated distributed storm water management system, in *Proceedings of the 1st ACM International Workshop on Cyber-Physical Systems for Smart Water Networks*, pp. 1–6 (2015)
6. U. Raza, A. Salam, On-site and external power transfer and energy harvesting in underground wireless. Electronics **9**(4), 681 (2020)
7. U. Raza, A. Salam, A survey on subsurface signal propagation. Smart Cities **3**(4), 1513–1561 (2020)
8. U. Raza, A. Salam, Wireless underground communications in sewer and stormwater overflow monitoring: Radio waves through soil and asphalt medium. Information **11**(2), 98 (2020)
9. U. Raza, A. Salam, Zenneck waves in decision agriculture: An empirical verification and application in EM-based underground wireless power transfer. Smart Cities **3**(2), 308–340 (2020)
10. Richmond Department of Public Services, 2017: RVA clean water plan. Technical report, United States Environmental Protection Agency, 2017
11. Richmond Department of Public Services, Annual report: City of Richmond public utilities. Technical report, United States Environmental Protection Agency, 2017
12. A. Salam, Pulses in the sand: Long range and high data rate communication techniques for next generation wireless underground networks. *ETD collection for University of Nebraska - Lincoln* (AAI10826112, 2018)
13. A. Salam, A comparison of path loss variations in soil using planar and dipole antennas, in *2019 IEEE International Symposium on Antennas and Propagation* (IEEE, Jul 2019)
14. A. Salam, Design of subsurface phased array antennas for digital agriculture applications, in *Proc. 2019 IEEE International Symposium on Phased Array Systems and Technology (IEEE Array 2019)*, Waltham, MA, USA, October 2019
15. A. Salam, A path loss model for through the soil wireless communications in digital agriculture, in *2019 IEEE International Symposium on Antennas and Propagation* (IEEE, Jul 2019)
16. A. Salam, Sensor-free underground soil sensing, in *ASA, CSSA and SSSA International Annual Meetings (2019)* (ASA-CSSA-SSSA, 2019)
17. A. Salam, Subsurface MIMO: A beamforming design in Internet of Underground Things for digital agriculture applications. J. Sensor Actuat. Netw. **8**(3), 41 (2019)
18. A. Salam, *Underground Environment Aware MIMO Design Using Transmit and Receive Beamforming in Internet of Underground Things* (Springer International Publishing, Cham, 2019), pp. 1–15
19. A. Salam, An underground radio wave propagation prediction model for digital agriculture. Information **10**(4), 147 (2019)
20. A. Salam, Underground soil sensing using subsurface radio wave propagation, in *5th Global Workshop on Proximal Soil Sensing*, COLUMBIA, MO, May 2019
21. A. Salam, *Internet of Things for Environmental Sustainability and Climate Change* (Springer International Publishing, Cham, 2020), pp. 33–69
22. A. Salam, *Internet of Things for Sustainability: Perspectives in Privacy, Cybersecurity, and Future Trends* (Springer International Publishing, Cham, 2020), pp. 299–327
23. A. Salam, *Internet of Things for Sustainable Community Development*, 1 edn. (Springer Nature, New York, 2020)
24. A. Salam, *Internet of Things for Sustainable Community Development: Introduction and Overview* (Springer International Publishing, Cham, 2020), pp. 1–31
25. A. Salam, *Internet of Things for Sustainable Forestry* (Springer International Publishing, Cham, 2020), pp. 147–181
26. A. Salam, *Internet of Things for Sustainable Human Health* (Springer International Publishing, Cham, 2020), pp. 217–242
27. A. Salam, *Internet of Things for Sustainable Mining* (Springer International Publishing, Cham, 2020), pp. 243–271

28. A. Salam, *Internet of Things for Water Sustainability* (Springer International Publishing, Cham, 2020), pp. 113–145
29. A. Salam, *Internet of Things in Agricultural Innovation and Security* (Springer International Publishing, Cham, 2020), pp. 71–112
30. A. Salam, *Internet of Things in Sustainable Energy Systems* (Springer International Publishing, Cham, 2020), pp. 183–216
31. A. Salam, *Internet of Things in Water Management and Treatment* (Springer International Publishing, Cham, 2020), pp. 273–298
32. A. Salam, A.D. Hoang, A. Meghna, D.R. Martin, G. Guzman, Y.H. Yoon, J. Carlson, J. Kramer, K. Yansi, M. Kelly, et al., The future of emerging IoT paradigms: Architectures and technologies (2019)
33. A. Salam, U. Karabiyik, A cooperative overlay approach at the physical layer of cognitive radio for digital agriculture, in *Third International Balkan Conference on Communications and Networking 2019 (BalkanCom'19)*, Skopje, Macedonia, the former Yugoslav Republic of, June 2019
34. A. Salam, U. Raza, *Autonomous Irrigation Management in Decision Agriculture* (Springer International Publishing, Cham, 2020), pp. 379–398
35. A. Salam, U. Raza, *Current Advances in Internet of Underground Things* (Springer International Publishing, Cham, 2020), pp. 321–356
36. A. Salam, U. Raza, *Decision Agriculture* (Springer International Publishing, Cham, 2020), pp. 357–378
37. A. Salam, U. Raza, *Electromagnetic Characteristics of the Soil* (Springer International Publishing, Cham, 2020), pp. 39–59
38. A. Salam, U. Raza, *Modulation Schemes and Connectivity in Wireless Underground Channel* (Springer International Publishing, Cham, 2020), pp. 125–166
39. A. Salam, U. Raza, On burial depth of underground antenna in soil horizons for decision agriculture, in *2020 International Conference on Internet of Things (ICIOT-2020)*, Honolulu, USA, September 2020
40. A. Salam, U. Raza, *Signals in the Soil*, 1st edn. (Springer Nature, New York, 2020)
41. A. Salam, U. Raza, *Signals in the Soil: An Introduction to Wireless Underground Communications* (Springer International Publishing, Cham, 2020), pp. 3–38
42. A. Salam, U. Raza, *Signals in the Soil: Subsurface Sensing* (Springer International Publishing, Cham, 2020), pp. 251–297
43. A. Salam, U. Raza, *Signals in the Soil: Underground Antennas* (Springer International Publishing, Cham, 2020), pp. 189–215
44. A. Salam, U. Raza, *Soil Moisture and Permittivity Estimation* (Springer International Publishing, Cham, 2020), pp. 299–317
45. A. Salam, U. Raza, *Underground Phased Arrays and Beamforming Applications* (Springer International Publishing, Cham, 2020), pp. 217–248
46. A. Salam, U. Raza, *Underground Wireless Channel Bandwidth and Capacity* (Springer International Publishing, Cham, 2020), pp. 167–188
47. A. Salam, U. Raza, *Variable Rate Applications in Decision Agriculture* (Springer International Publishing, Cham, 2020), pp. 399–423
48. A. Salam, U. Raza, *Wireless Underground Channel Modeling* (Springer International Publishing, Cham, 2020), pp. 61–121
49. A. Salam, S. Shah, Internet of Things in smart agriculture: Enabling technologies, in *2019 IEEE 5th World Forum on Internet of Things (WF-IoT) (WF-IoT 2019)*, Limerick, Ireland, April 2019
50. A. Salam, M.C. Vuran, Impacts of soil type and moisture on the capacity of multi-carrier modulation in Internet of Underground Things, in *Proc. of the 25th ICCCN 2016*, Waikoloa, Hawaii, USA, Aug 2016
51. A. Salam, M.C. Vuran, EM-based wireless underground sensor networks, in *Underground Sensing*, ed. by S. Pamukcu, L. Cheng (Academic Press, 2018), pp. 247–285

52. A. Salam, M.C. Vuran, Smart underground antenna arrays: A soil moisture adaptive beamforming approach, in *Proc. IEEE INFOCOM 2017*, Atlanta, USA, May 2017
53. A. Salam, M.C. Vuran, Wireless underground channel diversity reception with multiple antennas for Internet of Underground Things, in *Proc. IEEE ICC 2017*, Paris, France, May 2017
54. A. Salam, M.C. Vuran, X. Dong, C. Argyropoulos, S. Irmak, A theoretical model of underground dipole antennas for communications in Internet of Underground Things. IEEE Trans. Antenn. Propagat. **67**(6), 3996–4009 (2019)
55. A. Salam, M.C. Vuran, S. Irmak, Pulses in the sand: Impulse response analysis of wireless underground channel, in *The 35th Annual IEEE International Conference on Computer Communications (INFOCOM 2016)*, San Francisco, USA, April 2016
56. A. Salam, M.C. Vuran, S. Irmak, Towards Internet of Underground Things in smart lighting: A statistical model of wireless underground channel, in *Proc. 14th IEEE International Conference on Networking, Sensing and Control (IEEE ICNSC)*, Calabria, Italy, May 2017
57. A. Salam, M.C. Vuran, S. Irmak, Di-Sense: In situ real-time permittivity estimation and soil moisture sensing using wireless underground communications. Comput. Netw. **151**, 31–41 (2019)
58. A. Salam, M.C. Vuran, S. Irmak, A statistical impulse response model based on empirical characterization of wireless underground channel. IEEE Trans. Wirel. Commun. **19**, 5966–5981 (2020)
59. FJ. Tetazoog, Clean City, Green Waters, Lincoln City Journal (2022)
60. S. Temel, M.C. Vuran, M.M.R. Lunar, Z. Zhao, A. Salam, R.K. Faller, C. Stolle, Vehicle-to-barrier communication during real-world vehicle crash tests. Comput. Commun. **127**, 172–186 (2018)
61. United States Environmental Protection Agency, Report to congress implementation and enforcement of the combined sewer overflow control policy. Technical report, Office of Water, 2001
62. United States Environmental Protection Agency, Report to congress: Report to congress on implementation and enforcement of the CSO control policy. Technical report, Office of Water, 2001
63. United States Environmental Protection Agency, Report to congress: Impacts and control of CSOs and SSOs. Technical report, Office of Water, 2004
64. United States Environmental Protection Agency, Encouraging low impact development: Incentives can encourage adoption of lid practices in your community. Technical report, Office of Water, 2012
65. United States Environmental Protection Agency, Report to congress: Clean watersheds needs survey 2012. Technical report, Office of Water, 2012
66. United States Environmental Protection Agency, Greening CSO plans: Planning and modeling green infrastructure for combined sewer overflow (CSO) control. Technical report, Office of Water, 2014
67. M.C. Vuran, A. Salam, R. Wong, S. Irmak, Internet of Underground Things in precision agriculture: Architecture and technology aspects. Ad Hoc Netw. (2018). https://doi.org/10.1016/j.adhoc.2018.07.017
68. M.C. Vuran, A. Salam, R. Wong, S. Irmak, Internet of Underground Things: Sensing and communications on the field for precision agriculture, in *2018 IEEE 4th World Forum on Internet of Things (WF-IoT) (WF-IoT 2018)*, Singapore, February 2018

Chapter 6
Internet of Things in Sewer Monitoring

6.1 A New Paradigm: Sewer IOUT

The need of gathering real-time status information about sewer system has resulted in a different kind of Internet of Things (IoT): Internet of Underground Things (IOUT). It is different from IoT as it specifically focuses on gathering information from the underground applications like agriculture, pipeline assessment, landslide monitoring, border patrol, mining, oil and gas [2, 25, 26, 89, 96, 97, 103], etc. IOUT consists of sensor nodes which interconnects with the communication solutions to transfer information to and from the sewer system so that an appropriate decision can be made to prevent server overflows [33–36].

IOUT differs from IoT because of their special operating environment. It requires information from underground sewers about incoming and outgoing water flow and soil information and performs wireless communication and control actions. Given these unique requirements and operations, the traditional over-the-air communication may not be able to work. Therefore, a new paradigm of communication, i.e., wireless underground (UG) communications [3, 97], needs to be developed and optimized from underground applications. Underground communication has to be conducted through soil (in case of agriculture), concrete (in case of mining), and a combination of water and concrete (in case of sewer). In all scenarios, communication is done through either sensor nodes with in situ communication capabilities or stand-alone radios buried underground.

The chapter investigated the IOUT as a design solution for sewer and wastewater management system. Section 6.2 presents IOUT architecture with functionality of its components in detail. In Sects. 6.3 and 6.4, different sensing and communication solutions are discussed in detail. Finally, the chapter is concluded with some challenges that need to be addressed for the improvement of IOUT.

A. Salam, *Internet of Things in Smart Sewer and Drainage Systems*,
https://doi.org/10.1007/978-3-031-48109-3_6

6.2 IOUT Architecture

As discussed in the previous section, IOUT consists of heterogeneous devices that are interconnected to sense and gather information for sewer systems. This section discusses some of the desirable functionalities for sewer IOUT. Figure 6.1 shows the architecture of sewer IOUT.

- *In situ Sensing:* The major functionality of IOUT is its in situ sensing capabilities to have a localized knowledge of sewer systems. These sensors will gather information from the sewers, transmit the information to centralized cloud server, and trigger actuators based on the information received from the center server.
- *Wireless Communication in Challenging Environments:* The IOUT communication architecture consists of underground and above-ground components for communication. Communication modules are expected to take data from the sewers and transfer to cloud center. They are also responsible for implementing control actions and decision from cloud through actuator valves.
- *Inter-connection of Gray infrastructure, Sensors, Radios, and Cloud:* Since sewer IOUT has a diverse set of devices, it is desirable that the interconnection and communication between these devices should be seamless and automated.
- *Real-Time Decision-Making:* Underground sewer information gathered through sensor nodes and relayed through gateways to center cloud is always available

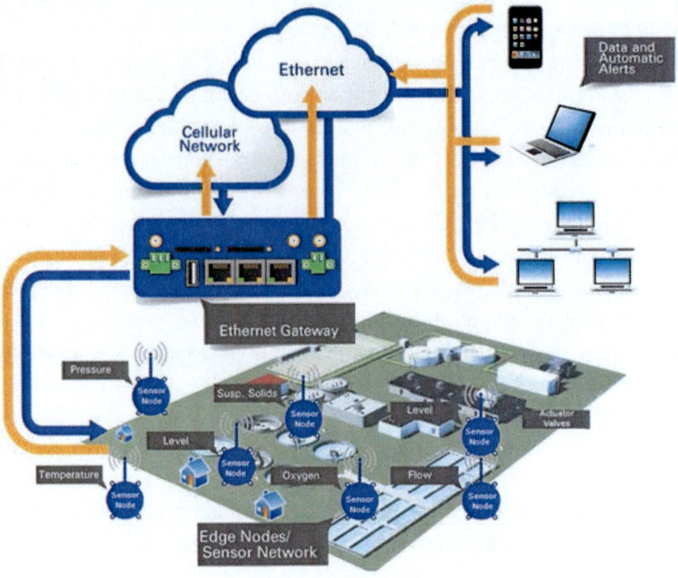

Fig. 6.1 IOUT Paradigm in Sewer Management System

for managers, decision support system, and decision-makers. This information is analyzed to make a real-time decision for optimized sewer management system.

- *Mobility:* Sewer IOUT will support both fixed desktop/computer devices and moving mobile devices. Moreover, it will also provide security in terms of short- and long-term backup of data.

An IOUT architecture is designed with consideration of desired functionalities and is shown in Fig. 6.1.

- *Underground Things (UTs):* A UT is an embedded sensor node having different sensors deployed on it. Multiple UTs will be used in the system where some of them will be buried aboveground and some of them underground. To protect UTs from harsh underground conditions and aboveground unpredictable weather, these UTs are protected by waterproof enclosure. UT sensor nodes gather information about pressure of water flow, water levels, flow speed of water, etc. These sensor nodes are fitted near sewer pipelines originating from consumer homes, outfalls near river, treatment plants, etc [38–45].
  These UTs also have embedded communication modules which are responsible for sending the sensor's data to gateways deployed in the field. Underground communication requires to communicate through soil and concrete, from sewer, and adjust parameters as per changing environment. Some existing communication methodologies include NFC, Wi-Fi, LoRa, LoRaWAN, ZigBee, Sigfox, Bluetooth, cellular, and satellite. A UT designed by academics [8, 24] has the ability to communicate over the distance of 100 meters, and a commercially designed UT can communicate over up to three times of this distance. Moreover, if communication is required over a very large distance, cellular and satellite technologies can also be used. Moreover, UT cost can be significantly lower, if a large of numbers of UTs are being deployed in the field [16].
- *Static Sinks:* Communication modules on UTs enable them to connect with the strategically places gateways aboveground. Sensor nodes are responsible for communicating underground parameters to aboveground communication gateways. Gateways further communicate the information to cloud network using networking protocols [46–51].
- *Mobile sinks* For larger fields, a moving sink node can be used to gather data and relay it to the longer distances. To that end, unmanned aerial vehicle (UAV) can be used as a mobile sink.
- *Cloud services* are required for a secure and permanent storage and backup of the data related to sewer events and analyze the data to make control decision most suitable with the data gathered from the field [52–57].

It is evident that sewer IOUT has a diverse communication and sensing devices and protocols. Such a diverse environment has its own challenges of interoperability and compatibility. Moreover, communication and sensing protocols further add to the complexity of the architecture. Therefore, creating a unified sewer IOUT architecture is a very challenging task.

6.3 Sensing

Sensing is a major functionality of sewer IOUT which initiates the process of IOUT sewer managements. Figure 6.2 shows different sensor types, e.g., water quality sensor, depth sensor, and soil moisture sensor that can be integrated on UT to provide sensed information. This section provides an overview of sensing technologies that can be used in sewer IOUT (Fig. 6.3).

Soil Moisture: Soil Moisture (SM) sensors have been used in agriculture sector for decades. Since sewer systems are made up of concrete, soil, gravel, etc., SM sensors can be very useful in gathering information about water accumulated in sewer infrastructure to keep a check on condition of the system. Traditionally, soil moisture values are measured manually using handheld devices. Some of the famous methods of SM measurement are explained below:

- Gravimetric sampling is used to calculate the volumetric water content in the soil by using the ratio of soil mass in dry state to the soil mass in wet state. This is a manual sampling method and samples are dried by heating them in oven [13].
- Resistive sensors [22] take electrical conductivity of water into account while measuring soil moisture. It calculates the resistive changes in the soil in response to VWC of the soil. However, this method requires the proper calibration of the sensors to get accurate readings [63–71].
- Capacitive sensors take the capacitance of water into account while measuring soil moisture. They are more expensive than resistive sensors but are also more accurate than resistive sensors [72].

Fig. 6.2 Sensing and control equipment for IOUT Sewer

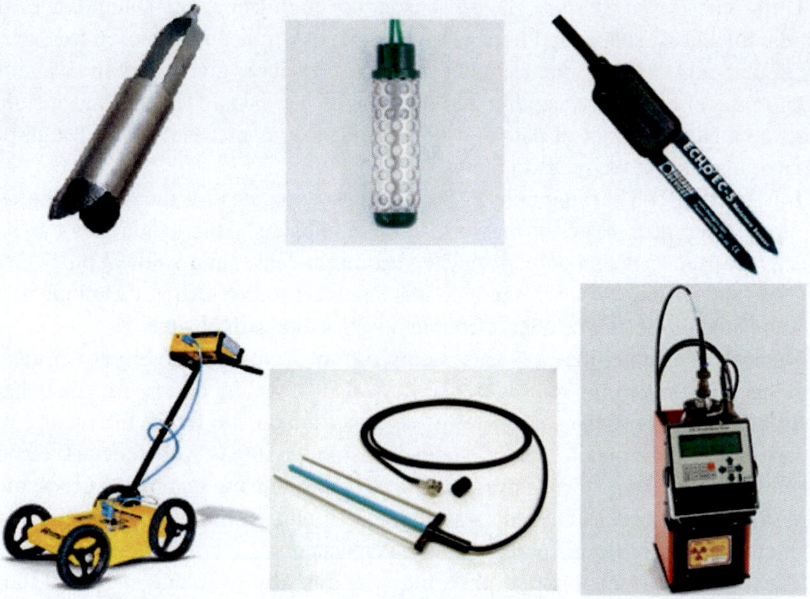

Fig. 6.3 Soil moisture sensors. Top row: Gravimetric [13], resistive (Watermark) [22], and capacitance [18]. Bottom row: GPR [19], TDR [37], and neutron probe [14]

- Ground Penetrating Radar (GPR) [1, 19] takes absorption and reflection phenomena of electromagnetic waves into account while measuring SM. This method is more popular in the application where soil moisture needs to be measured from the near surfaces, e.g., 10 cm [73–81].
- Neutron scattering probes [9, 14] and gauges measure the change in neutron flux density of soil due to water. It is considered as one of the most accurate methods of measuring soil moisture, and however, it requires a specific license to be implemented.

  In addition to the above-mentioned methods, other methods include time-domain reflectometry (TDR) [37], gamma ray attenuation [12], and frequency-domain reflectometry (FDR) [90].

Weather and Environmental Sensing: Weather and environmental factors like temperature, rainfall, wind speed, solar emission, and humidity are key to predict whether a combined sewer overflow (CSO) will occur or not. Such sensors have already been introduced by John Deere [23]. This information helps in developing an informed and controlled IOUT sewer management system. An existing weather system, mesoscale network (MesoNet), comes with the functionality of sensing aforementioned weather and environmental parameters and that too over a large geographical area. Therefore, MesoNet can be combined with existing sewer IOUT solutions to get real-time weather and environmental information [82–87].

Ultrasonic-Based Sensors: These sensors use high-frequency sound waves that test the thickness, shape, and defects of the sewer. A transducer is used to generate ultrasound pulses. Each sound signal coming at transducer gives amplitude and time taken to travel between inception and retrieval of the signal [15]. However, it does generate a large amount of data and non-homogeneous materials (e.g., rough pipe surface) may affect the measurements.

Infra-Red (IR) Thermography: Thermography operates by measuring the temperature differences between the surface of the objects. This techniques has been effectively used to detect defects, like deterioration, leaks, and voids in pipes [102]. It is safe and radiation-free. Moreover, it is an efficient area testing technique which measures the properties of a large area in a very short span of time.

Remote Sensing: Since sewer systems do not have a favorable condition for working and operating, remote sensing is also a viable option for such harsh environment. Remote sensing makes use of electromagnetic waves interacting with under-sewer environment. Such methods measure the intensity of reflected electromagnetic waves from sewer infrastructure. Remote sensing has been successful in precision agriculture, satellite data fusion, and condition monitoring, and it would be interesting to see how it performs in sewer systems.

These sensing technologies pave the way for advancement in sewer IOUT. The easily available and inexpensive sensing equipment with communication capabilities is key to the integration into IOUT ecosystem. In addition to sensing, wireless communication is also a major component of sewer IOUT and is key to IOUT. A robust connectivity in field is as important as sensing the information from the field. Therefore, wireless communication to connect such a diverse set of devices plays a key role in realization of sewer IOUT.

6.4 Wireless Connectivity

There are two types of connectivity solutions required for sewer IOUT: one for communicating within the field and the other for connecting outside the field with cloud.

6.4.1 In-Field Communications

In-field communication is required to connect UTs with the gateways deployed in the field. Most of the commercially available communication solution works perform OTA communication, whereas, for sewer, the major requirement is to communicate underground from sewers. For the short-range communication within the field, between UT and gateway, a license-free solution, e.g., Bluetooth, DASH 7, Zigbee, etc., can be used. Recently, Federal Communication Commission (FCC)

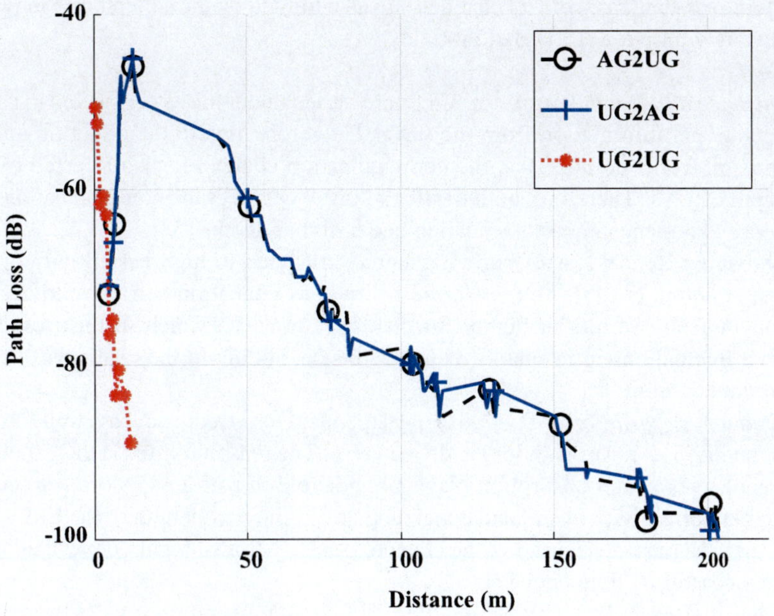

Fig. 6.4 Communication from soil

has allowed the use of restricted bands for farm usage [11], which reduces the interference from licensed devices.

UG Communications: UG communication involves both underground, i.e., between UTs, and aboveground, i.e., between UTs and gateways. Moreover, the construction material (soil and concrete) of sewer interface also affects aboveground communication and requires a detailed analysis and modeling. Figure 6.4 models the path loss for an agricultural field experiment [8], and however, a similar behavior can be expected from sewer underground communication because of having a similar surface between aboveground and underground. It is observed that the practical underground communication distance is limited by 12 m and aboveground can have up to 200 m of communication.

UG communication is impacted by several factors, and some of those factors are explained below:

- *Soil Moisture:* Sewer material is composed of both concrete and soil. Therefore, for accurately understanding the communication through sewer, it is imperative to understand how both material will response to EM signals propagating through the soil.

  Soil conductivity depends upon the soil moisture, and hence, soil water content impacts EM signals. Dry soil and near-saturation level soil have conductivity between 10^{-4} and 10^{-5} Si/m. However, the dielectric constant for dry soil is between 2 and 6 and that of near-saturation level is between 5 and 15 [94].

Moreover, date rates of UG channels are also limited by the coherence bandwidth and are within few 100 kHz [58–60].

- *Concrete:*
- *Distance and Depth Variations:* UTSs are buried under the sewer system to gather operational information from the sewer. Therefore, due to the effect of soil–air and soil–concrete interface, the communication channel is also affected by the depth of UTs. Therefore, both depth and distance affect the communication. UT nodes experience higher attenuation under higher depths [58].
- *Antennas:* Return loss of buried antenna varies due to high permittivity of soil and concrete [7, 95]. Moreover, the continuous change in soil permittivity and soil moisture results in highly fluctuating return loss which in turn results in high fluctuations in resonant frequency, hence, adding to the challenges of UG communication.
- *Frequency Variations:* Path loss is dependent on frequency as well. Lower frequency has lower path loss and vice versa. The reason for this is that, at higher frequencies, water absorption plays a major role in path loss. Moreover, due to higher permittivity of soil and concrete than air, the wavelength of the EM waves is also decreased. Therefore, the channel capacity is also highly dependent upon the operational frequency [7].
- *Lateral Waves:* As mentioned earlier, UG communication occurs between UTs and exhibits one of the three paths, i.e., Direct, Lateral, or Reflected [8, 59, 61]. Among these lateral wave experiences the lowest attenuation and is crucial to the success of UG communications [92].
- *Recent Advances in UG communication:* There has been an active research going on in the field of UG communication due to its applications. Some of these research efforts include characterization of UG channel, cross-layer communication, achieving high data rates, long-distance communication, and environment-aware solutions.

  One such effort is made by Salam et al. [61, 62] to completely characterize the UG communication channel. They performed experiments in a greenhouse testbed and took more than 1500 measurements to analyze the statistical model of UG communication. To that end, they studied the effect of soil moisture and soil texture, UG channel impulse response, validated direct, reflected, and lateral wave components. It also shows that dynamic techniques are required for UG communication which can adapt to the changing soil moisture values.

  After analyzing the impulse response of a UG channel, the authors in [58] develop techniques to achieve high data rate communication. Moreover, they also analyze the effect of soil type, moisture on system capacity, and underground antenna. This analysis forms the base for the development of wireless UG channel diversity reception schemes and multi-carrier modulation [60].

  Another technique, adaptive combining (AC-LDR), dynamically removes the unwanted interference. Depending upon the distance of LDR receiver, AC-LDR selects either L-wave or D-wave component, whichever is the strongest. Since R-wave is considered to be the weakest component, it is ignored as it may cause further performance degradation. The authors in [60] investigate the performance of different modulation schemes in terms of BER and were able to get much better SNR as compared to their existing counterpart methods. Sensing in sewer IOUT is highly dependent upon UG efficient UG communication, and therefore, it is desirable to have reliable communication and high data rate.

Soil moisture adaptive beamforming (SMABF) is developed by the authors of [59] using UG transmitter antenna array and an omnidirectional receiving antenna. The authors maximized L-wave component by carefully directing the energy of antenna at an optimum angel [98, 99].

Magnetic Induction (MI) and Acoustic UG Communications: MI-based UG communication has a very fast decay rate of receive signal strength (RSS) [27, 91], and therefore, it is very difficult to get high data rate or long-range communication using MI. Moreover, the wavelength of MI-based communication is very large. Due to these reasons, MI-based communication methods are not useful for sewer IOUT.

Underwater communication and UG communication are also similar in terms of harsh environment. However, EM waves have a much higher rate of degradation in water, and therefore, an acoustic technology is used instead of EM waves.

Underground to UAV Communications: The unmanned vehicles have emerged as an effective way of sensing, monitoring, and communicating in the field [17, 32]. Before advancement in UAVs, satellite imagery was used as an alternative for monitoring. However, UAVs are not only fast but also an inexpensive way of field mapping. However, UG to UAV communication also faces challenges like low range, limited flight times, a specific operator exhibiting a special skill set, and licenses for flying drone.

Low-Power Wide-Area Networks (LPWANs): Two major desirable characteristics of IOUT are long-range connectivity and longevity. Low-Power Wide-Area Network (LPWAN) aims to achieve longevity in the network by conserving energy, hence, forcing sensor to operate for longer periods of time without even changing them for many years. Therefore, LPWAN is the most suitable technology in this regard [30]. LPWAN applications are expected to last for longer periods of time (several years) and are suitable for the application that sends smaller packets and do not require transfer of data for longer time periods. The following are some of the related technologies for LPWAN.

(1) LoRa: Long-Range Wide-Area Network (LoRaWAN) uses one-hop star topology to conserve energy. It can transfer the packet size between the range of 0.290 and 50 kbps over the distance of 45 km in rural areas [6]. LoraWAN uses the chord spread spectrum modulation technique, adaptive data rate technique, and gateway used for decoding received signals to achieve reliable and long-range communication.

(2) Sigfox: Sigfox [88] uses ultra narrow band (UBR) modulation, has a range of 45 km in rural and 12 km in urban area, and provides the data rate of 250 kbps. It uses an unlicensed spectrum between 868 MHz and 902 MHz for communication. However, the data rate is also limited by regulations [104]. Sigfox can be a promising communication solution for sewer IOUT.

(3) On-Ramp/Ingenu: On-Ramp is a PHY/MAC layer technical standard which is developed by IEEE 802.15.4k [21]. It has a communication range of 15 km and operates between the frequency range of 902 and 928 MHz in the unlicensed spectrum. It is more suitable for sewer IOUTs spanning over large geographical areas.

(4) NB-IoT: NB-IoT [100] is a physical layer protocol developed by 3GPP LTE. It is more suitable for low data rate application. It operates in the licensed spectrum. Moreover, it supports multiple types of operations, i.e., in-band broadband and stand-alone operation. It is suitable for prolonged, cost-efficient, and long-distance sewer IOUTs.

(5) Extended Coverage GSM IoT: EC-GSM-IoT [10] has a battery life of up to 10 years. It operates in two GSM bands: 800 MHz to 900 MHz and 1800 to 1900 MHz. EC-GSM has an inexpensive equipment, prolonged operational life, and extended coverage area which makes it suitable for sewer IOUTs.

In addition to the above-mentioned LPWAN technologies, other notable technologies are Weightless [101], Ingenu [21], Platanus [31], and NWave [29].

Wireless PAN/LAN: Wireless LAN/PAN has high data rate and low latency for IOUT. Since LPWAN are more focused on providing extended communication range, these functionalities are missing from LPWAN. Hence, wireless LAN/PAN is a suitable solution for sewer IOUT where the main priorities are data rate and low latency. Wireless technologies are briefly explained as follows:

(1) Bluetooth: Bluetooth [5] is a standard developed by Special Interest Group (SIG). It can transmit data up to a distance of 100 m and has a bandwidth of 25 MHz. There is another version of the standard, Bluetooth Smart, which consumes low energy. Bluetooth is also being used in the development of various low-energy sensors [4].

(2) ZigBee: Zigbee falls under the application and network layers of the protocol stack. It can communicate for up to the distance range of 10–30 meters and has a bandwidth of 1 MHz. This technology can be very useful in sewer IOUT due to its simple operation.

(3) Thread: Thread [93] is another IEEE 802.15.4 MAC/PHY protocol. Its main features include being secure, energy efficient, and the ability to connect up to 250 devices. It can be used in sewer IOUT for communication between UTs and gateways within the field for longer time periods.

(4) Wi-Fi: Wi-fi, a physical layer protocol, is also known as IEEE 802.11 standard [20]. It supports the highest data rate of up to 1 Gbps. It utilizes multiple ISM bands and has a bandwidth of up to 160 MHz. In [28], the researchers have developed wireless sensor nodes that connect with monitoring system using Wi-Fi technology.

Cellular Technology in IOUT :Advancement in IOUT technologies has increased the demand for cellular and broadband connectivity. Cellular connectivity is the major problem in rural areas. Since there is a glut of data being generated in the field, the lack of high-speed connectivity hinders the adaption of smart technologies in the rural areas. Moreover, a huge amount of expenditure is required to develop infrastructure for the fast cellular connectivity in rural areas. Moreover, it is also important for the IOUT devices to be compatible with the cellular technologies. In addition to compatibility, these devices must also be energy efficient or at least have an alternate energy source for prolonged operation. An alternative approach could be to have two types of connectivity solutions: low-powered devices within the field can be connected through in-field communication, and data can be transferred to the cloud through advanced gateways operating on external power source.

Energy Consumption in IOUT: Minimal energy consumption is the primary requirement of sewer IOUTs. A low-powered sewer IOUT can enable sensor nodes with minimal energy requirement to operate for longer periods of time without replacing batteries.

6.4.2 Cloud and Big Data in Sewer IOUT

Sensors in the sewer generate a glut of data which cannot be stored locally because of limited calculation and processing power. To that end, data are processed on cloud solutions and stored in central databases. Databases are kept shared or private depending upon the privacy policies. Moreover, cloud services provide visualization of the data, processing power, and scalability for bigger sewer network. On contrary, without any storage constraints, the base station combines the sensor's data and automated weather data and takes decision to open or control the actuator valve depending on weather data. Therefore, there should be a way to control the sewer equipment remotely. IOUT and sewer system integration open up new avenues for various stakeholders, e.g., consumers, municipalities, water treatment companies, etc.

Another important challenge is to combine the diverse set of devices in sewer IOUT ecosystem. With such a diverse set of devices and heterogeneous set of protocols, reliable data transfer becomes more important than ever. In future, the researchers are encouraged to address the challenge by developing standardized interfaces to provide interoperability as well as seamless connection between different components of IOUT ecosystem.

Even a minor change in a sewer IOUT may result in the generation of a large amount of data. Such a large amount of data makes it difficult to extract meaningful information from it. This process is very crucial to all the stakeholders to make correct decision to get rate on investment and wide adaption of IOUT paradigm. Therefore, it is imperative to develop big data analytics in sewer IOUT.

In addition to data analysis and processing power, it is also important to analyze the effect of adopting IOUT on labor and energy consumption cost. City municipalities are the major stakeholders of the sewer IOUT since they are the one who allocate big budgets for the infrastructure. Therefore, it is important to show them the advantages of big data analytics in the sewer domain without burdening them with the unnecessary detail of data. This will help getting more cities to adopt this practice and realize the cost saving and benefits for the consumers.

6.5 Research Challenges

This section lists out some of the major challenges in implementing IOUT in large scale.

1. Since IOUT solution is going to be deployed in the large geographical sewer field, spanning over the cities, it is very important to have a low-cost and sustainable IOUT devices which can bear the harsh environment of underground sewers.
2. Improving UT operation will also add complexity to the devices, thus making them energy hungry and reducing their battery life. Therefore, improving UT with sustainable energy demand is highly important.
3. Multitude of diverse sensor devices makes it difficult to integrate with even diverse communication equipment and protocols. Therefore, standard IOUT protocols are required to make this integration seamless.
4. Security is another important aspect of IOUTs. Most of the stakeholders are sensitive about the data and want to store them on secure and private databases. Therefore, it is important to provide a secure storage and communication method to gain trust of the stakeholders.
5. Variation in temperature and weather may affect the operational parameters of the devices, e.g., security of devices under-sewer during heavy rainfall or storm. Therefore, these natural calamities and variations should be taken into consideration while implementing IOUT.
6. Traditional OTA link layer and network layer protocols must be tweaked to make them suitable for underground environment, and scalable, reliable, and high data rate networking protocol must be developed.

References

1. V.I. Adamchuk, J.W. Hummel, M.T. Morgan, S.K. Upadhyaya. On-the-go soil sensors for precision agriculture. Comput. Electron. Agric. **44**(1), 71–91 (2004)
2. M.A. Akkaş, Channel modeling of wireless sensor networks in oil, in *Wireless Personal Communications* (2017), pp. 1–19
3. I.F. Akyildiz, E.P. Stuntebeck, Wireless underground sensor networks: Research challenges. Ad Hoc Networks Journal (Elsevier) **4**, 669–686 (2006)
4. J. Bjarnason, *Evaluation of Bluetooth Low Energy in Agriculture Environments* (2017)
5. Bluetooth. https://www.bluetooth.com/bluetooth-technology
6. J. de Carvalho Silva, J. JPC Rodrigues, A.M. Alberti, P. Solic, A.L.L. Aquino, Lorawan—a low power wan protocol for internet of things: A review and opportunities, in *2017 2nd International Multidisciplinary Conference on Computer and Energy Science (SpliTech)* (IEEE, New York, 2017), pp. 1–6
7. X. Dong, M.C. Vuran, Impacts of soil moisture on cognitive radio underground networks, in *Proceedings of the IEEE BlackSeaCom*, Batumi, Georgia (2013)
8. X. Dong, M.C. Vuran, S. Irmak, Autonomous precision agriculture through integration of wireless underground sensor networks with center pivot irrigation systems. Ad Hoc Netw. **11**(7), 1975–1987 (2013)

9. S.R. Evett, J.L. Steiner, Precision of neutron scattering and capacitance type soil water content gauges from field calibration. Soil Sci. Soc. Am. J. **59**(4), 961–968 (1995)
10. Extended coverage—GSM—Internet of Things (EC-GSM-IoT). https://www.gsma.com/iot/extended-coverage-gsm-internet-of-things-ec-gsm-iot/
11. FCC order no. da 16–307 dated: Mar 24, 2016. https://apps.fcc.gov/edocs_public/attachmatch/DA-16-307A1.pdf
12. H. Ferguson, W.H. Gardner, Water content measurement in soil columns by gamma ray absorption. Soil Sci. Soc. Am. J. **26**(1), 11–14 (1962)
13. H.D. Foth, *Fundamentals of Soil Science*, 8 edn. (Wiley, New York, 1990)
14. T.E. Franz, A. Wahbi, M. Vreugdenhil, G. Weltin, L. Heng, M. Oismueller, P. Strauss, G. Dercon, D. Desilets, Using cosmic-ray neutron probes to monitor landscape scale soil water content in mixed land use agricultural systems, in *Applied and Environmental Soil Science*, vol. 2016 (2016)
15. E. Ginzel, Ultrasonic inspection 2: training for nondestructive testing variables affecting test results. NDT. net **4**(6) (1999)
16. J. Gutierrez, J.F. Villa-Medina, A. Nieto-Garibay, M.A. Porta-Gandara, Automated irrigation system using a wireless sensor network and GPRS module. IEEE Trans. Instrum. Meas. **63**(1), 166–176 (2014)
17. S.R. Herwitz, L.F. Johnson, S.E. Dunagan, R.G. Higgins, D.V. Sullivan, J. Zheng, B.M. Lobitz, J.G. Leung, B.A. Gallmeyer, M. Aoyagi, R.E. Slye, J.A. Brass, Imaging from an unmanned aerial vehicle: agricultural surveillance and decision support. Comput. Electron. Agric. **44**(1), 49–61 (2004)
18. Honeywell. https://sensing.honeywell.com/sensors/
19. J.A. Huisman, S.S. Hubbard, J.D. Redman, A.P. Annan, Measuring soil water content with ground penetrating radar. Vadose Zone J. **2**(4), 476–491 (2003)
20. IEEE 802.11 technical standard—wireless local area networks. http://http://www.ieee802.org/11/
21. IEEE 802.15.4 technical standard. https://www.ieee802.org/15/pub/TG4.html
22. Irrometer, Soil moisture sensor. http://www.irrometer.com/sensors.html
23. John Deere field connect. https://www.deere.com/en/technology-products/precision-ag-technology/field-and-water-management/field-connect/
24. Y. Kim, R.G. Evans, W.M. Iversen, Remote sensing and control of an irrigation system using a distributed wireless sensor network. IEEE Trans. Instrum. Meas. **57**(7), 1379–1387 (2008)
25. S. Kisseleff, I.F. Akyildiz, W.H. Gerstacker, Digital signal transmission in magnetic induction based wireless underground sensor networks. IEEE Trans. Commun. **63**(6), 2300–2311 (2015)
26. A. Konda, A. Rau, M.A. Stoller, J.M. Taylor, A. Salam, G.A. Pribil, C. Argyropoulos, S.A. Morin, Soft microreactors for the deposition of conductive metallic traces on planar, embossed, and curved surfaces. Adv. Funct. Mater. **28**(40), 1803020 (2018)
27. S. Lin, I.F. Akyildiz, P. Wang, Z. Sun, Distributed cross-layer protocol design for magnetic induction communication in wireless underground sensor networks. IEEE Trans. Wirel. Commun. **14**(7), 4006–4019 (2015)
28. G.R. Mendez, M.A. Md Yunus, S.C. Mukhopadhyay, A Wi-Fi based smart wireless sensor network for an agricultural environment, in *2011 Fifth International Conference on Sensing Technology* (2011), pp. 405–410
29. NWave. https://www.nwave.io/
30. J. Petäjäjärvi, K. Mikhaylov, M. Hämäläinen, J. Iinatti, Evaluation of lora LPWAN technology for remote health and wellbeing monitoring, in *2016 10th International Symposium on Medical Information and Communication Technology (ISMICT)* (IEEE, New York, 2016), pp. 1–5
31. Platanus. https://www.m2comm.co/front-page/technology/lan-platanus/
32. J. Primicerio, S.F. Di Gennaro, E. Fiorillo, L. Genesio, E. Lugato, A. Matese, F.P. Vaccari, A flexible unmanned aerial vehicle for precision agriculture. Precis. Agric. **13**(4), 517–523 (2012)

33. U. Raza, A. Salam, On-site and external power transfer and energy harvesting in underground wireless. Electronics **9**(4), 681 (2020)
34. U. Raza, A. Salam, A survey on subsurface signal propagation. Smart Cities **3**(4), 1513–1561 (2020)
35. U. Raza, A. Salam, Wireless underground communications in sewer and stormwater overflow monitoring: radio waves through soil and asphalt medium. Information **11**(2), 98 (2020)
36. U. Raza, A. Salam, Zenneck waves in decision agriculture: An empirical verification and application in EM-based underground wireless power transfer. Smart Cities **3**(2), 308–340 (2020)
37. K. Roth, R. Schulin, H. Flühler, W. Attinger, Calibration of time domain reflectometry for water content measurement using a composite dielectric approach. Water Resour. Res. **26**(10), 2267–2273 (1990)
38. A. Salam, Pulses in the sand: Long range and high data rate communication techniques for next generation wireless underground networks. *ETD collection for University of Nebraska—Lincoln (AAI10826112)* (2018)
39. A. Salam, A comparison of path loss variations in soil using planar and dipole antennas, in *2019 IEEE International Symposium on Antennas and Propagation* (IEEE, New York, 2019)
40. A. Salam, Design of subsurface phased array antennas for digital agriculture applications, in *Proceedings of the 2019 IEEE International Symposium on Phased Array Systems and Technology (IEEE Array 2019)*, Waltham, MA, USA (2019)
41. A. Salam, A path loss model for through the soil wireless communications in digital agriculture, in *2019 IEEE International Symposium on Antennas and Propagation* (IEEE, New York, 2019)
42. A. Salam, Sensor-free underground soil sensing, in *ASA, CSSA and SSSA International Annual Meetings (2019)*. ASA-CSSA-SSSA (2019)
43. A. Salam, Subsurface MIMO: a beamforming design in internet of underground things for digital agriculture applications. J. Sens. Actuator Netw. **8**(3), 41 (2019)
44. A. Salam, *Underground Environment Aware MIMO Design Using Transmit and Receive Beamforming in Internet of Underground Things* (Springer International Publishing, Cham, 2019), pp. 1–15
45. A. Salam, An underground radio wave propagation prediction model for digital agriculture. Information **10**(4), 147 (2019)
46. A. Salam, Underground soil sensing using subsurface radio wave propagation, in *5th Global Workshop on Proximal Soil Sensing*, COLUMBIA, MO (2019)
47. A. Salam, *Internet of Things for Environmental Sustainability and Climate Change* (Springer International Publishing, Cham, 2020), pp. 33–69
48. A. Salam, *Internet of Things for Sustainability: Perspectives in Privacy, Cybersecurity, and Future Trends* (Springer International Publishing, Cham, 2020), pp. 299–327
49. A. Salam, *Internet of Things for Sustainable Community Development*, 1 edn. (Springer Nature, New York, 2020)
50. A. Salam, *Internet of Things for Sustainable Community Development: Introduction and Overview* (Springer International Publishing, Cham, 2020), pp. 1–31
51. A. Salam, *Internet of Things for Sustainable Forestry* (Springer International Publishing, Cham, 2020), pp. 147–181
52. A. Salam, *Internet of Things for Sustainable Human Health* (Springer International Publishing, Cham, 2020), pp. 217–242
53. A. Salam, *Internet of Things for Sustainable Mining* (Springer International Publishing, Cham, 2020), pp. 243–271
54. A. Salam, *Internet of Things for Water Sustainability* (Springer International Publishing, Cham, 2020), pp. 113–145
55. A. Salam, *Internet of Things in Agricultural Innovation and Security* (Springer International Publishing, Cham, 2020), pp. 71–112
56. A. Salam, *Internet of Things in Sustainable Energy Systems* (Springer International Publishing, Cham, 2020), pp. 183–216

57. A. Salam, *Internet of Things in Water Management and Treatment* (Springer International Publishing, Cham, 2020), pp. 273–298
58. A. Salam, M.C. Vuran, Impacts of soil type and moisture on the capacity of multi-carrier modulation in internet of underground things, in *Proceedings of the 25th ICCCN 2016*, Waikoloa, Hawaii, USA (2016)
59. A. Salam, M.C. Vuran, Smart underground antenna arrays: a soil moisture adaptive beam-forming approach. in *Proceedings of the IEEE INFOCOM 2017*, Atlanta, USA (2017)
60. A. Salam, M.C. Vuran, Wireless underground channel diversity reception with multiple antennas for internet of underground things, in *Proceedings of the IEEE ICC 2017*, Paris, France (2017)
61. A. Salam, M.C. Vuran, S. Irmak, Pulses in the sand: Impulse response analysis of wireless underground channel, in *Proceedings of the IEEE INFOCOM 2016*, San Francisco, USA (2016)
62. A. Salam, M.C. Vuran, S. Irmak, Towards internet of underground things in smart lighting: a statistical model of wireless underground channel, in *Proceedings of the 14th IEEE International Conference on Networking, Sensing and Control (IEEE ICNSC)*, Calabria, Italy (2017)
63. A. Salam, M.C. Vuran, X. Dong, C. Argyropoulos, S. Irmak, A theoretical model of underground dipole antennas for communications in internet of underground things. IEEE Trans. Antennas Propag. **67**(6), 3996–4009 (2019)
64. A. Salam, M.C. Vuran, S. Irmak, A statistical impulse response model based on empirical characterization of wireless underground channel. IEEE Trans. Wirel. Commun. **19**(9), 5966–5981 (2020)
65. A. Salam, A.D. Hoang, A. Meghna, D.R. Martin, G. Guzman, Y.H. Yoon, J. Carlson, J. Kramer, K. Yansi, M. Kelly, et al., The future of emerging IoT paradigms: Architectures and technologies (2019)
66. A. Salam, U. Karabiyik, A cooperative overlay approach at the physical layer of cognitive radio for digital agriculture, in *Third International Balkan Conference on Communications and Networking 2019 (BalkanCom'19)*. Skopje, Macedonia, the former Yugoslav Republic of (2019)
67. A. Salam, U. Raza, *Autonomous Irrigation Management in Decision Agriculture* (Springer International Publishing, Cham, 2020), pp. 379–398
68. A. Salam, U. Raza, *Current Advances in Internet of Underground Things* (Springer International Publishing, Cham, 2020), pp. 321–356
69. A. Salam, U. Raza, *Decision Agriculture* (Springer International Publishing, Cham, 2020), pp. 357–378
70. A. Salam, U. Raza, *Electromagnetic Characteristics of the Soil* (Springer International Publishing, Cham, 2020), pp. 39–59
71. A. Salam, U. Raza, *Modulation Schemes and Connectivity in Wireless Underground Channel* (Springer International Publishing, Cham, 2020), pp. 125–166
72. A. Salam, U. Raza, On burial depth of underground antenna in soil horizons for decision agriculture, in *2020 International Conference on Internet of Things (ICIOT-2020)*, Honolulu, USA (2020)
73. A. Salam, U. Raza, *Signals in the Soil*, 1 edn. (Springer Nature, Berlin, 2020)
74. A. Salam, U. Raza, *Signals in the Soil: An Introduction to Wireless Underground Communications* (Springer International Publishing, Cham, 2020), pp. 3–38
75. A. Salam, U. Raza, *Signals in the Soil: Subsurface Sensing* (Springer International Publishing, Cham, 2020), pp. 251–297
76. A. Salam, U. Raza, *Signals in the Soil: Underground Antennas* (Springer International Publishing, Cham, 2020), pp. 189–215
77. A. Salam, U. Raza, *Soil Moisture and Permittivity Estimation* (Springer International Publishing, Cham, 2020), pp. 299–317
78. A. Salam, U. Raza, *Underground Phased Arrays and Beamforming Applications* (Springer International Publishing, Cham, 2020), pp. 217–248

79. A. Salam. U. Raza, *Underground Wireless Channel Bandwidth and Capacity* (Springer International Publishing, Cham, 2020), pp. 167–188
80. A. Salam, U. Raza, *Variable Rate Applications in Decision Agriculture* (Springer International Publishing, Cham, 2020), pp. 399–423
81. A. Salam, U. Raza, *Wireless Underground Channel Modeling* (Springer International Publishing, Cham, 2020), pp. 61–121
82. A. Salam, S. Shah, Internet of things in smart agriculture: Enabling technologies, in *2019 IEEE 5th World Forum on Internet of Things (WF-IoT) (WF-IoT 2019)*, Limerick, Ireland (2019)
83. A. Salam, M.C. Vuran, EM-Based wireless underground sensor networks, in *Underground Sensing*, ed. by S. Pamukcu, L. Cheng (Academic Press, New York, 2018), pp. 247–285
84. A. Salam, M.C. Vuran, S. Irmak, Di-sense: In situ real-time permittivity estimation and soil moisture sensing using wireless underground communications. Comput. Netw. **151**, 31–41 (2019)
85. A. Salam, M.C. Vuran, Smart underground antenna arrays: A soil moisture adaptive beamforming approach, in *Proceedings of the IEEE INFOCOM 2017*, Atlanta, USA (2017)
86. A. Salam, M.C. Vuran, S. Irmak, Pulses in the sand: impulse response analysis of wireless underground channel. in *The 35th Annual IEEE International Conference on Computer Communications (INFOCOM 2016)*, San Francisco, USA (2016)
87. A. Salam, M.C. Vuran, S. Irmak, Towards internet of underground things in smart lighting: a statistical model of wireless underground channel. in *Proceedings of the 14th IEEE International Conference on Networking, Sensing and Control (IEEE ICNSC)*, Calabria, Italy (2017)
88. Sigfox. https://www.sigfox.com/en
89. A.R. Silva, M.C. Vuran, $(CPS)^2$: integration of center pivot systems with wireless underground sensor networks for autonomous precision agriculture, in *Proceedings of ACM/IEEE International Conference on Cyber-Physical Systems*, Stockholm, Sweden (2010), pp. 79–88
90. W. Skierucha, A. Wilczek, A FDR sensor for measuring complex soil dielectric permittivity in the 10–500 MHz frequency range. Sensors **10**(4), 3314–3329 (2010)
91. X. Tan, Z. Sun, I.F. Akyildiz, Wireless underground sensor networks: MI-based communication systems for underground applications. IEEE Antennas Propag. Mag. **57**(4), 74–87 (2015)
92. S. Temel, M.C. Vuran, M.M.R. Lunar, Z. Zhao, A. Salam, R.K. Faller, C. Stolle, Vehicle-to-barrier communication during real-world vehicle crash tests. Comput. Commun. **127**, 172–186 (2018)
93. Thread. https://www.threadgroup.org/technology/ourtechnology
94. F.T. Ulaby, D.G. Long, *Microwave Radar and Radiometric Remote Sensing* (University of Michigan Press, Michigan, 2014)
95. M. Vuran, X. Dong, D. Anthony, Antenna for wireless underground communication, December 27 2016. US Patent 9,532,118
96. M.C. Vuran, I.F. Akyildiz, Cross-layer packet size optimization for wireless terrestrial, underwater, and underground sensor networks, in *IEEE INFOCOM 2008. The 27th Conference on Computer Communications* (2008), pp. 226–230
97. M.C. Vuran, I.F. Akyildiz, Channel model and analysis for wireless underground sensor networks in soil medium. Phys. Commun. **3**(4), 245–254 (2010)
98. M.C. Vuran, A. Salam, R. Wong, S. Irmak, Internet of underground things in precision agriculture: architecture and technology aspects. Ad Hoc Networks **81**, 160–173 (2018)
99. M.C. Vuran, A. Salam, R. Wong, S. Irmak, Internet of underground things: sensing and communications on the field for precision agriculture, in *2018 IEEE 4th World Forum on Internet of Things (WF-IoT) (WF-IoT 2018)*, Singapore (2018)
100. Y.-P.E. Wang, X. Lin, A. Adhikary, A. Grovlen, Y. Sui, Y. Blankenship, J. Bergman, H.S. Razaghi, A primer on 3gpp narrowband internet of things. IEEE Commun. Mag. **55**(3), 117–123 (2017)
101. Weightless. http://www.weightless.org/

102. G.J. Weil, R.J. Graf, L.M. Forister, Remote sensing pipeline rehabilitation methodologies based upon the utilization of infrared thermography, in *Urban Drainage Rehabilitation Programs and Techniques: Selected Papers on Urban Drainage Rehabilitation from 1988–1993 Water Resource Planning and Management Division Conference Sessions* (ASCE, New York, 1994), pp. 173–181
103. S.U. Yoon, L. Cheng, E. Ghazanfari, S. Pamukcu, M.T. Suleiman, A radio propagation model for wireless underground sensor networks, in *2011 IEEE Global Telecommunications Conference—GLOBECOM 2011* (2011), pp. 1–5
104. X. Zhang, A. Andreyev, C. Zumpf, M.C. Negri, S. Guha, M. Ghosh, Thoreau: a subterranean wireless sensing network for agriculture and the environment, in *2017 IEEE Conference on Computer Communications Workshops (INFOCOM WKSHPS)* (2017), pp. 78–84

Chapter 7
Antenna as Sensor in Drainage and Sewer Systems

7.1 Introduction

The characteristics and impacts of electromagnetic waves in soil medium are of vital significance in many research areas of smart agriculture [11, 39–52]. In this regard, two major properties play a very important role in propagation of electromagnetic radiations: first the composition of medium and electric field distribution in the medium. Knowing the medium composition parameters gives a bigger picture of how EM field effects the medium. Moreover, electric field distribution helps in understanding the EM radiation of the medium under some effect, e.g., soil moisture. In practical applications, the above-mentioned quantities, i.e., medium composition and electric field distribution, are not possible to determine using mathematical equations. Instead, a suitable antenna must be designed to determine these properties. To that end, a linear antenna can be used as a probe to determine these properties in agricultural IoT settings [8–10, 22–31, 38, 58].

The antenna can be mounted on some vehicle, driven antenna, or penetrated in the ground, receiving probe, to sense required properties. The probes have different applications in many areas. For example, in geo- and bio-physics, driven probes can be used to determine the properties and composition of the medium like soil, rock, ice, biological tissues, etc. Similarly, receiving antenna can measure the electromagnetic pulse penetration after nuclear explosion. They are also used to determine tissue damage in biological specimen that has been exposed to radiation.

In this chapter, we have investigated the use of antenna for sensor-free soil moisture measurements with applications in digital agriculture where measuring soil moisture is a difficult task. The following constraints hinders the process:

- Soil is very heterogeneous in nature with many components mixed with it.
- The spatial and temporal variability of soil also poses a major challenge.
- Sensors are deployed deep in soil and are not easily accessible.
- It is messy to work with the soil and autonomously verify it.

A. Salam, *Internet of Things in Smart Sewer and Drainage Systems*,
https://doi.org/10.1007/978-3-031-48109-3_7

- The accuracy of the data produced by soil moisture sensors might not always be correct.

It is possible to measure the soil moisture at any one point or location in the field, and however, it is a very labor-intensive, time-taking, and costly procedure. Therefore, it is desirable to measure the soil moisture of a larger patch of the field instead of a single point along with the immediate production of results. This can be done by the radio waves. The radio wave propagating toward the ground attenuates because of the two major reasons: (1) conductivity of the ground and (2) dielectric constant of the Earth. Both properties depend upon the soil moisture value, and therefore, the power and intensity of the radio waves can be determined by measuring the value of soil moisture. This research proposes to explore the feasibility of measuring soil moisture through radio wave propagation and identify the challenges that may be encountered during the process [7, 12–21, 32–37, 53–57, 59, 61, 62].

The section aims to present the current soil moisture measuring techniques. The very detailed discussion is not given at this stage, and however, sufficient information is provided to understand the crux and basic principal of each method so that the readers can easily find the related information in regard to the research presented in this book. The next section discusses the theoretical concepts of radio wave propagation such as surface and space wave, skin depth, and the relation between the dielectric constant and radio wave propagation. The theoretical discussion is helpful in understanding the measuring parameters. There are many soil moisture measuring methods that are developed on the basis of physical principles. We categorize them into electrical resistivity, nuclear, gravimetric, hygrometric, capacitance, radiation-based, thermal conductivity, tensiometry, and miscellaneous. Almost all fall into the category of measuring soil moisture in a point or a small area and are time-intensive methods. Their application differs due to size, reliability, accuracy, speed, cost, and the type of measurements. The basic information about each method is given in the following, which is required to understand the principle, salient features, and where the readers can get more detailed information about a respective method.

Hygrometric

It is a well-known fact that the atmosphere relative humidity is related to the moisture of porous material under equilibrium conditions. Hygrometric is a method to measure soil moisture under the relative humidity just above the ground level. A detailed discussion on humidity sensors (also known as hydrometers) is given in [47]. Electrical resistance hygrometers follow the principle that change in electrical resistance can result in the change of soil moisture and electrolytic process. The materials used for the process are salt and aluminum oxide. Salt disassociates in water and lowers the electrical resistance present between both electrodes. Aluminum oxide acts as a dielectric film for capacitors and its resistance and dielectric constant value changes with the change in humidity.

Our electrolytic system has two winded platinum wires. These wires are coated with partially hydrated phosphorus pentoxide (P_2OS). These wires are kept inside an insulated tube. The gas flows above the wire along with the direct current passing through the wires simultaneously. P_2OS absorbs the water vapors. The direct current breaks water into H_2 and O_2 components. As soon as both, water absorbed by P_2OS and water broken by the electrolysis, reach equilibrium, relative humidity is calculated using current flow, pressure, and temperature.

Another such system, similar to described above, uses a wire wound inside a tubular wick highly filled with a hygroscopic salt, i.e., lithium chloride (LiCl). It applies alternating voltage at two ends of the wounded wires. Coated LiCl becomes conductor as soon as it absorbs humidity and can pass the current generating the heat directly proportional to the current flow. Equilibrium is reached without losing or gaining water from the layer. Humidity is determined from the temperature associated with the equilibrium. There are two well-known types of hygrometers: (1) capacitor transducers with dielectric and (2) microwave refractometers with a cavity. The dielectric constant of the former changes with the change in moisture. The cavity of the later has resonant frequency varying with the dielectric constant and humidity.

A piezoelectric absorption-based hygrometer uses a quartz crystal with a resonant frequency dependent upon the mass of the crystal. The crystal is coated with the hygroscopic material. The mass varies as the crystal absorbs moisture and humidity is calculated from the change in resonant frequency. Optical methods are also used for the humidity sensors. Reference [8] uses an optical instrument operating in an infrared region. The system monitors two selected frequencies and the ratio of their amplitude is a function of water vapor concentration.

Many materials, e.g., dacron thread, human hair, white oak, scotch pinstripes, and some ceramic materials, vary dimensionally with humidity and are used to build hygrometers. The relative humidity is measured through the change in dimensions of these materials. A dew-point hygrometer is dependent upon the temperature. It is simple and inexpensive but has a very slow response time. The dew appears, on the surface of a cooled object, at some certain temperature. This temperature can be related with the partial water pressure of the gas sample for measurement. Another simple method of making hygrometer is using the psychrometric instruments. It uses the temperature between the dry and wet bulbs and also uses this information to determine the relative humidity using a psychometric formula.

Electrical Resistivity
Soil resistance is dependent upon the moisture present in it. Therefore, this fact can be used to calculate a soil moisture measure by using simple resistance measurements. However, there are some issues in measuring resistance: (1) resistance due to the contact of electrode and (2) resistance that may occur due to the presence of ionized salts. Therefore, two different measurements are taken to address this issue. Sample resistivity measures the resistivity between the electrodes directly in contact with the soil sample [20]. The second measurement is known as sensor resistivity. It measures the resistance of sensor when the sensor reaches the moisture equilibrium.

Capacitance
The change in soil moisture affects the capacitance of the soil. It is due to the fact that the dielectric constant of soil (2.6) differs from the dielectric of water (80). The

difference in dielectric value can result in measurement errors. Like the resistance method, there are two methods of measuring soil moisture: (1) soil capacitance and (2) sensor capacitance. The capacitance method is discussed in detail in [13].

Nuclear

Neutron scattering work on the principle of measuring slowing the of the neutron emitted by fast-neutron source. Soil moisture contains the low-weight hydrogen atoms. Fast moving neutrons are slowed after the collision with low-weight hydrogen atoms in soil moisture. Therefore, the number of slow neutrons is proportional to the hydrogen atoms which are due to water content in the soil. This method is also known as volumetric measurement. Some measurement constraints to this method are sample density, sample composition, sample roughness, and sample homogeneity. There exist an extensive research and literature on how these constraints can be minimized, and however, it is not possible to cover all those here. Another nuclear method to determine soil moisture is gamma ray transmission. Gamma ray transmission can measure soil moisture within half-inch layer of the soil.

Gravimetric

A total of five steps are involved in measuring soil moisture using the gravimetric method. Those five steps are as follows:

1. Collection of soil sample
2. Weigh the sample
3. Drying the sample to remove moisture
4. Weighing the dried sample again
5. Calculate moisture content using difference of weights

This is the direct method to measure the soil moisture and is used as a calibration for other techniques. Various literature provide application of this technique [28]. One of the popular methods to remove moisture from the sample is thermal extraction. Thermal extraction is done using different techniques such as oven drying, distillation, freeze drying, heating in oil, and alcohol burning. Chemical extraction is another method to remove moisture which uses alcohol, Karl Fischer reagent, and calcium carbide (hydride). Mechanical extraction uses mechanical pressing to remove water.

Radiation

The radiation method uses two types of radiations to measure soil moisture: mechanical and electromagnetic (EM) radiations. Mechanical radiation uses high and low frequency ranges of ultrasonic waves and peneterometer, respectively. The effect of ultrasonic waves on soil moisture is also of significant nature. The peneterometer works on the basic principle that resistance from the soil varies due to soil moisture.

EM radiations are classified into four different categories: radio waves, nuclear magnetic resonance, infrared, and microwave. Radio and microwaves use the same principle of absorption. Only the difference between both lies in the frequency range and the equipment being used. In a nuclear magnetic resonance technique, a soil sample is simultaneously placed in the fixed and varying magnetic field. At some specific frequencies, the energy absorption is increased under the changing magnetic field. This happens because of nuclear magnetic resonance and water molecules are measured through hydrogen nuclei spectra. Finally, infrared reflection involves the measurement of reflection from the exposed surface. This reflection is proportional to the amount of soil moisture in the solid surface. However, this is not the most accurate method because it is the surface measurement and will vary if the surface is not homogeneous.

Tensiometry
Tensiometer is a device which measures soil moisture using capillary tension induced by water in the soil. There are two types of tensiometer: (1) porous cup tensiometer and (2) permeable membrane tensiometer. The permeable membrane uses the osmosis principal for flow and related the capillary tension to soil moisture.

Thermal Conductivity
Soil moisture is related to thermal conductivity. The change in thermal conductivity is proportional to the change in soil moisture. This fact is used to measure the soil moisture by rising the temperature of material kept at some known distance from the heat source. Researchers have developed a technique to apply the above principal using a thermistor giving dual functionality of heat element and temperature indicator [39].

Miscellaneous
The volume of the water expands by a large amount of evaporation. Therefore, a soil sample with a small volume can be heated in a container causing the water to evaporate. This will result in the rise of container pressure and sample moisture can be related to this container pressure [39]. Another way to measure soil moisture is through colorimeter. It contains cobalt chloride which after mixing with soil moisture changes its color as per the amount of moisture in the soil.

7.2 Antenna as Sensor: The System Model

The distribution of current on a highly conducting linear antenna immersed in a material medium is a function of the dimensions of the antenna, frequency of operation, and the three scalar constitutive parameters that describe the linear isotropic medium: namely, the real effective conductivity ae, the real effective permittivity (dielectric constant), and the permeability, which is assumed to be real and the same as for free space for the materials considered here. When the antenna is used as a sensor, the constitutive parameters of the medium must be determined from a measurement of the current on the antenna. The most convenient place on the antenna to measure the current is at the driving point. If the effect of the junction between the antenna and the feeding transmission line is small, this is approximately equivalent to measuring the driving-point admittance or impedance. In principle, a measurement of the distribution of current along the entire length of the antenna would yield more information about the surrounding medium. However, apparatus for this purpose is more complex, and there is difficulty with a movable sensor on the antenna when the surrounding material is a solid. In many instances a single measurement of the resonance or terminal admittance or impedance can be used to determine accurately the constitutive parameters of the surrounding medium. The conductors of both antennas are alike; the insulated probe is formed by adding a concentric sheath with effective permittivity. In this chapter we consider the effective resonant frequency in the linear, homogeneous, isotropic medium surrounding the antenna in soil.

Equation (7.1) gives the formula for determining the frequency of antenna resonance:

$$AR_{f_{min}} = F(\max(RL_f)) \, , \tag{7.1}$$

where the frequency of the maximum return loss is represented by $AR_{f_{min}}$. The value of $AR_{f_{min}}$ is independent of the separation distance between sender and receiver antennas. $AR_{f_{min}}$ is ineffective of distance because of antenna gains causing AR to be calculated with antenna return loss with minimal noise effect due to narrow band. Soil factor β_s is measured using the following equation:

$$\beta_s = AR_{f_{min}}/\upsilon \, , \tag{7.2}$$

where the resonant frequency of the antenna is given by υ. It is important to mention that this υ is for free space. Moreover, the value for wavelength at the υ is calculated using the following equation:

$$\Lambda = c/\upsilon \, , \tag{7.3}$$

where c represents the speed of light and Λ represents the wavelength at resonant frequency υ. The following equation calculates the relative permittivity of the soil using the values of wavelength and soil factor:

$$E_{rel} = \frac{1}{(\beta_s \times \Lambda)^2} \, . \tag{7.4}$$

The operational frequency of the system where the input impedance of antenna becomes equal to pure resistance is known as resonant frequency υ. This can be represented as [1]

$$\text{Imp}_a^u|_{f=f_r} = \text{Imp}_r = \text{Res}_a \tag{7.5}$$

where return loss is maximum such that

$$f_{RL} = \max(RL_{dB}). \tag{7.6}$$

These computations are for resonant antenna, and however, it would be the same for longer antennas with zero input reactance. The zero reactance will occur at anti-resonant length. The resonant antenna method can be used in all soil types. The purpose is to calculate the electrical properties of medium. A monopole antenna is between the interface of medium or air and is made resonant using two different method. The first method is to vary the length l with a non-variable fixed frequency, and the other is to vary frequency with fixed length. Any of the stated methods can be used to successfully establishing the resonance, and then electrical properties of the medium are measured.

Soil has various properties, e.g., moisture, permittivity, bulk density, etc. Among these properties, moisture and permittivity are interlinked in a way that soil permittivity is only dependent on soil moisture, and however, both soil permittivity and moisture are not dependent on other properties, i.e., frequency, texture, and bulk density of the soil [5, 60]. Relative permittivity for water and dry soil is 80 and 3, respectively. Using (7.4) for soil permittivity, the soil moisture is calculated using the following equation from [5, 60]:

$$VWC(\%) = \frac{E_{rel} - 3}{0.77} + 14.97\,. \tag{7.7}$$

Permittivity is a measurement of how a dielectric medium effects and effected by an electric field. It is a physical quantity which is calculated by the extent to which a material is polarized when exposed to the electric field reducing the field inside the material. To summarize, permittivity can be defined as the ability of the material to transmit an electric field. The imaginary permittivity depends upon the molecular relaxations, frequency, and electrical conductivity. Molecular relaxation is an energy loss that does not participate in dipole moment. EC is the indicator of salinity of soil. Other dielectrics are listed below:

1. Dielectric-Type Liquid Materials
2. NR-NC Non-relaxing Non-conducting Water
3. Isopropyl Alcohol
4. Castor Oil/IPA
5. R-NC Relaxing Non-conducting 1-Propanol, Glycerol
6. NR-C Non-relaxing Conducting Saline Water
7. Saturated Sand with Saline Water
8. R-C Relaxing and Conducting Saline Glycerol

Other important characteristics are outlined below:

- Real Dielectric Constant of water is 80.
- Real Dielectric Constant of dry soil is 2.
- Real Dielectric Permittivity is proportional to changes in soil moisture and can be empirically evaluated.

The estimation of soil moisture is highly dependent upon permittivity. Therefore, it is important to quantify the permittivity values. To that end, during production, QC data are collected in more than 10 steps. The dielectric of production units is compared with the standard dielectric to simulate soil response.

7.3 Empirical Measurements

In this work the measurement parameters are investigated along with the values of frequency, transmitter–receiver (T–R) antenna separation distance, and height of the antenna. Optimal values were chosen in accordance with the discussion on theory. Given all these parameters and considering various other factors, instruments were designed. Here we present the actual measurement obtained through the equipment and parameters discussed in the precious section. Measurement parameters like reflections, antenna height, vegetation, and temperature are investigated. Moreover, effective width, i.e., the width of the area effecting the propagation wave, is also examined. Finally, the relation between the received power of the signal and that of soil moisture value is examined. Due to harsh winter conditions, there was a limited time available to collect the data. Therefore, quantitative evaluation of the method could not be completed. However, the investigation will provide a basis to explore viability of the method and ground work for the additional research on the method. Figure 7.1 classifies soil into a textural triangle and Table 7.1 shows the exact component distribution of each soil. Soil is made up of clay, sand, and silt particles. The soil type is determined by the percentage of amount of these particles in a soil. For that purpose, a soil textural triangle is used to clearly show what a soil is made up of. Resonant frequency is analyzed with reference to textural triangle. For this analysis, volumetric water content is kept with the range of **5** 40% and OTA antenna of **433** megahertz is used. Accordingly, the resultant resonant frequencies are mapped over the textural triangle. This triangle can be used as a reference for the future prediction of resonant frequencies in underground environment. However, information about soil type and soil moisture level is needed for the prediction.

Antenna return loss occurs due to impedance mismatch and is calculated as $RL_{dB} = 20\log_{10}(Z_s + Z_a)/(Z_s - Z_a)$. Before starting the experiment, return loss of each transmitter–receiver (T–R) pair is measured at different depths and distance values. Experiments were conducted to understand the soil effect on communication. The water potential varied between 0 and 50 CB.

At 0 CB, i.e., full saturation, resonant frequency, and corresponding return loss at that resonant frequency value are very different for both types of the soil given all depths. Resonant frequency is 282 MHz in sandy soil and **211** MHz in silt loam soil under the depth of **10** cm. The difference is large because the relative permittivity is directly related to the amount of water present in a particular soil [2]. Silt loam soil can hold more water as compared to that of sandy soil and, hence, has a lower resonant frequency.

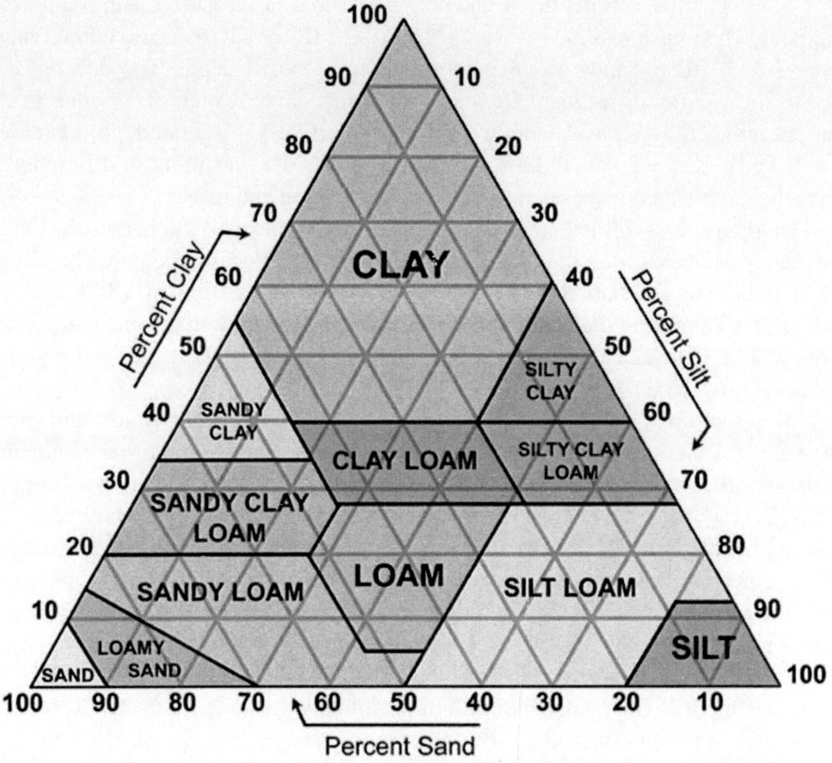

Fig. 7.1 Soil textural triangle

Table 7.1 Percentage of clay, sand, and silt in different soil types

Soil type	Clay %	Sand %	Silt %
Silt	9	10	81
Silt loam	20	30	50
Silty clay loam	35	15	50
Silty clay	55	10	35
Loam	20	50	30
Clay loam	35	42	23
Clay	60	38	2
Sandy clay	40	58	2
Sandy clay loam	30	65	5
Sandy loam	15	65	20
Loamy sand	10	80	10
Sand	8	91	1

The effect of burial depth on antenna return loss is measured using empirical methods. The return loss ranges from **−15** to −20 dB for silt loam soil and it ranges from **3** to 10 dB in sandy soil depending upon the burial depth. The difference is due to the change in resonant frequency and impedance which occurs due to the current induced by wave reflection at soil–air interface [3]. As the depth increases, the distance between soil and air interface also increases resulting in difference in intensity of reflected waves due to high soil absorption and phase.

The return loss of antenna also depends upon the soil moisture. Resonant frequency decreases with the increase in soil moisture. We also compared the return loss for silt loam soil buried at **10** cm. It compares the loss at 0 and 40 CB. It is worth noting that increasing the water potential causes the increase in resonant frequency from **212** MHz to **222** MHz. Hence, it validates the theoretical results presented in the previous period [4].

The constitutive parameters of fresh water, solutions of salt water, and moist sand are also measured with an arrangement discussed above. The resonant antenna method was used to measure the relative permittivity of lake water as a function of frequency of the range 12 MHz–90 MHz. The results of these measurements are also applicable to the soil. Also shown is the low-frequency relative permittivity of 78.5 for pure water at a temperature of 25ř C.7. The relative permittivity measured with the resonant antenna method differs by a maximum of 4% from this value. It is also worth noting that the electrical parameters for five saltwater solutions (NaCl) measured using the resonant antenna method at a fixed frequency of 28.01 MHz are also available [6]. The relative permittivity is roughly the same for the five solutions.

In these different methods of measurement, however, the applicability of each method is restricted by limitations from material properties and frequency range being used in a specific method. For the resonant antenna method, the resonant length is the main limitation. It is important for all parameters to match the method. For example, for any soil under consideration, a resonant length should be there for a particular frequency of operation. To further elaborate this point, it is important to note that the resonant length l for a dipole with value of $D = 10.0$. These calculations were made using different frequencies of different geological media having varying constituent parameters. Different soils had different values of soil moisture, and therefore, an average soil moisture value is taken. The frequency ranges of VHF band, i.e., **30**–300 MHz, resonant length is between few centimeters and few meters. These values remain constant for most of the soils considered for calculation. Therefore, these results advocate the efficacy of resonant antenna method for measuring the electrical properties of medium in VHF band using reasonable antenna sizes. Since the antenna length is inversely proportional to frequencies, i.e., the smaller antenna length for high frequencies, this method can further be extended to lower frequencies if there are experiment facilities to facilitate antennas with long lengths. In this analysis, experiments are performed to compare the value of $\frac{f_{rs}}{f_{ro}}$ at frequencies of **433** MHz. Different soil types, i.e., sandy and silty clay loam, are used under varying depths. Results are shown in the next section.

7.4 Results

In this section results of VWC and resonant frequency are presented for different soils. The results of silty clay soil are discussed first.

7.4.1 Silty Clay

It is observed, Fig. 7.2, that resonant frequency is dependent on depth and ratio and is also changing as compared to OTA. Two resonant frequencies, one from theoretical model and the other from permittivity antenna, were measured. The difference between both types of resonant frequencies, δ, is plotted. Results are shown for both soil types under different depths and moisture level at frequencies of **433** MHz and **915** MHz. It can be seen that δ is inversely proportional to soil moisture levels and directly proportional to frequencies. Its value increased by **10** MHz–**15** MHz when the frequency is increased from **433** MHz to **915** MHz. Therefore, it can be concluded that, if only permittivity is used, the operation frequency may fall out of the antenna bandwidth region. There will be no or

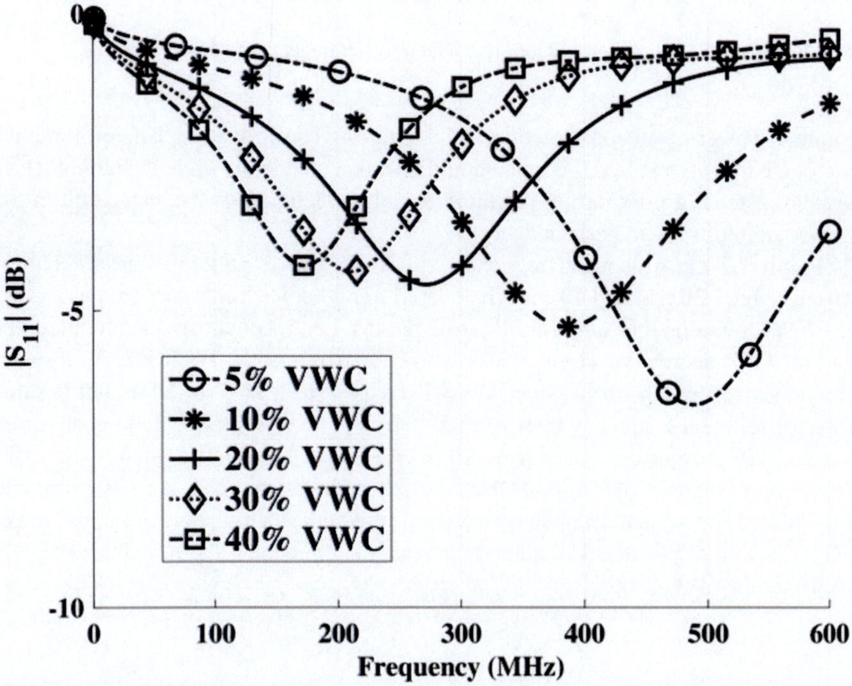

Fig. 7.2 Silty Clay: resonant frequency with return loss for different soil moisture levels

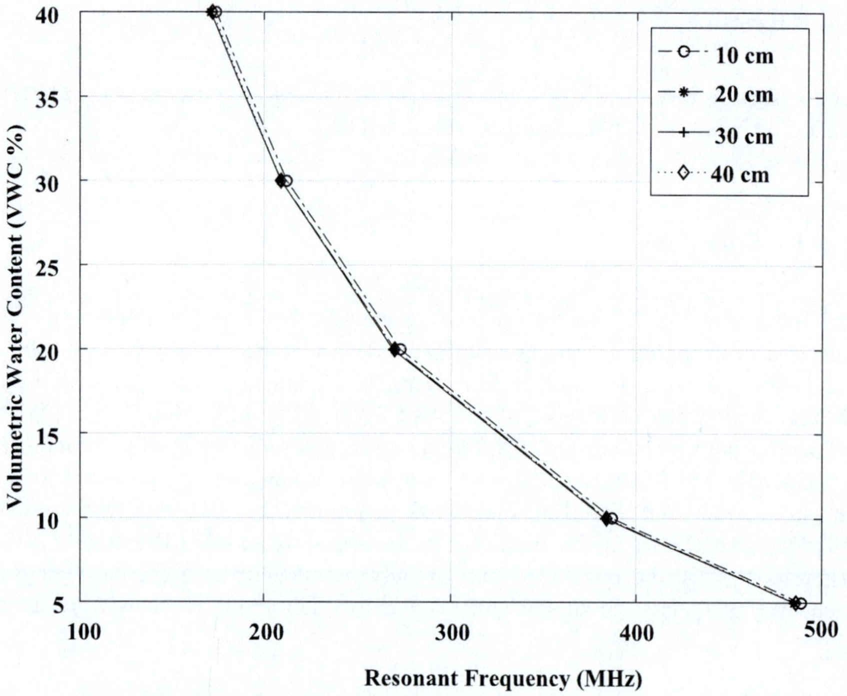

Fig. 7.3 Silty Clay: change in soil moisture with resonant frequency at different depths

minimal power of antenna to communicate through soil medium. Therefore, other factors, e.g., antenna burial depth, should also be considered while designing smart system instead of considering permittivity only which, otherwise, can result in the degradation of system performance.

Figure 7.3 plots resonant frequency with VWC values under the burial depth of **10** cm, **20** cm, **30** cm, and **40** cm. The overall trend between resonant frequency and the VWC remains the same for all burial depths, i.e., resonant frequency increases as the VWC decreases. The biggest change (approximately **100** MHz) in resonant frequency can be observed when VWC decreases from **20** % to **10** %. During this interval resonant frequency goes from **270** to 380MHz. A similar trend is observed when VWC further decreases from **10** % to **5** % and resonant frequency increases from approximately **380** to 480MHz. For other ranges of VWC, i.e., **40**–30% and **30**–20%, the resonant frequency increases gradually. This shows that for lower VWC in soil, the resonant frequency increases significantly and it increases slowly as the soil's VWC increases.

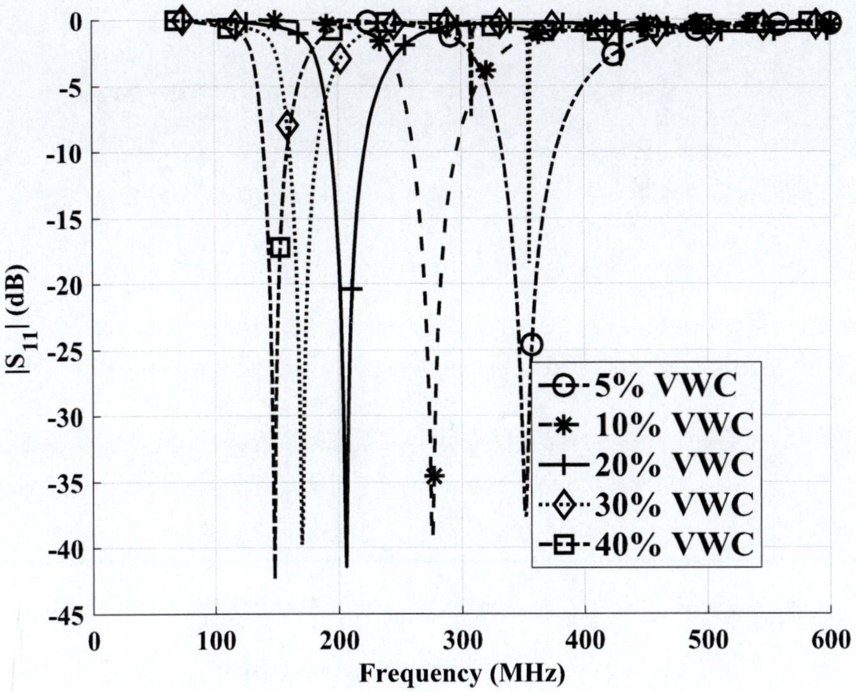

Fig. 7.4 Silt: resonant frequency with return loss for different soil moisture levels

7.4.2 *Silt*

Figure 7.4 plots the return loss with frequency under different VWC values, i.e., **5**–40%. It is important to note that higher values of return loss are more preferable. The experiment shows that the overall trend between frequency and the return loss remains the same for all VWC values, i.e., return loss is higher at higher values of frequency. The lowest value of return loss is observed at **40** % VWC and **140** MHz. Return loss is highest at **5** % VWC and **350** MHz. This shows that the signal strength will increase at lower VWC values and higher frequencies and it will decrease at higher VWC values and low frequencies.

Figure 7.5 compares the response of resonant frequency to the change in soil water content, i.e., VWC. The varying burial depth of antenna is also used to observe the effect of burial depth on resonant frequency. Antenna was buried at **10** cm, **20** cm, **30** cm, and **40** cm. It can be seen from the figure that resonant frequency has inverse relation with VWC, i.e., resonant frequency increases as the water content in the soil decreases. For example, the value of resonant frequency is approximately **160** MHz at **40** % VWC and **360** MHz at **5** %. Therefore, resonant frequency is increased by approximately **200** MHz between **40** and 5% VWC. Moreover, a similar trend is observed for the burial depths as well, i.e., resonant

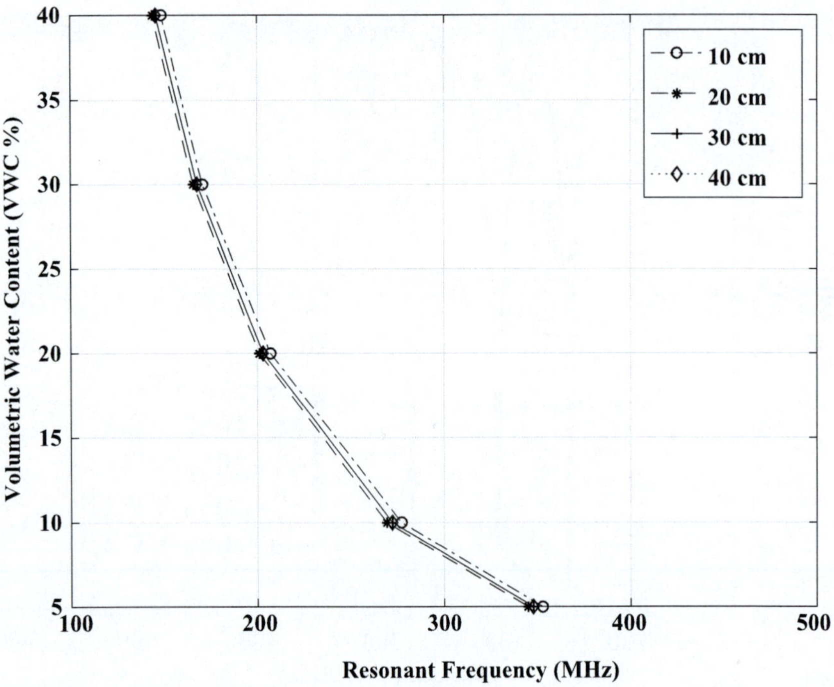

Fig. 7.5 Silt: change in soil moisture with resonant frequency at different depths

frequency increases with the decrease in burial depth. However, this increase is not that significant. Therefore, it can be concluded that VWC of soil is more significant than the burial depth of antenna in determining the resonant frequency.

7.4.3 Silty Clay Loam

Figure 7.6 shows the graph plotted for experiments performed under varying VWC percentages of soil. Return loss was observed for different frequencies and changing VWCs. In simple terms, return loss represents the signal strength. Therefore, higher values of return loss means the signal is powerful. To that end, an inverse relation is observed between frequency and return loss. Return loss is approximately **−5** dB at **180** MHz, whereas it is decreased to approximately **−8** dB at **500** MHz. On contrary, there is a direct relation between values of VWC and return loss, i.e., return loss decreases with decrease in VWC and vice versa.

Figure 7.7 plots resonant frequency with VWC values for different burial depths of **10** cm, **20** cm, **30** cm, and **40** cm. It can be seen that, for all burial depths, resonant frequency increases as VWC decreases. A significant increase is observed in frequency when VWC decreases from **20** to 10%, i.e., approximately **130** MHz.

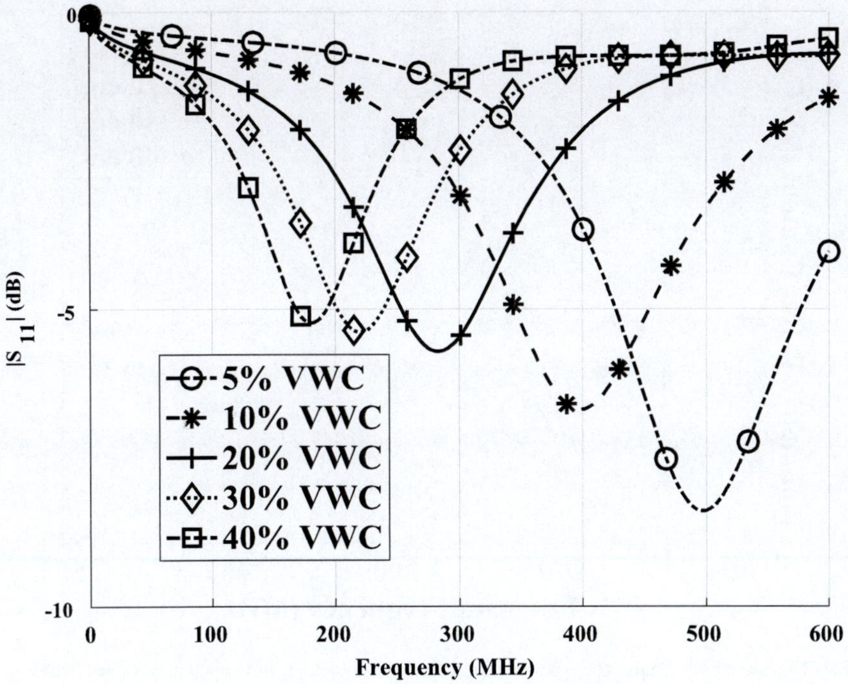

Fig. 7.6 Silty Clay Loam: resonant frequency with return loss for different soil moisture levels

The minimum change in frequency is observed between VWC range of **40** and 30%. This behavior shows that as the water content in the soil increases, the values of resonant frequency tend to decrease and vice versa.

7.4.4 Sandy Loam

Figure 7.8 compares the response of resonant frequency to the change in soil water content, i.e., VWC. The varying burial depth of antenna is also used to observe the effect of burial depth on resonant frequency. Antenna was buried at **10** cm, **20** cm, **30** cm, and **40** cm. It can be seen that resonant frequency has inverse relation with VWC, i.e., resonant frequency increases as the water content in the soil decreases. For example, the value of resonant frequency is approximately **170** MHz at **40** % VWC and **400** MHz at **5** %. Therefore, resonant frequency is increased by approximately **230** MHz between **40** and 5% VWC. Moreover, a similar trend is observed for the burial depths as well, i.e., resonant frequency increases with the decrease in burial depth. However, the increase is not that significant. Therefore, it can be concluded that VWC of soil is more significant than the burial depth of antenna in determining the resonant frequency.

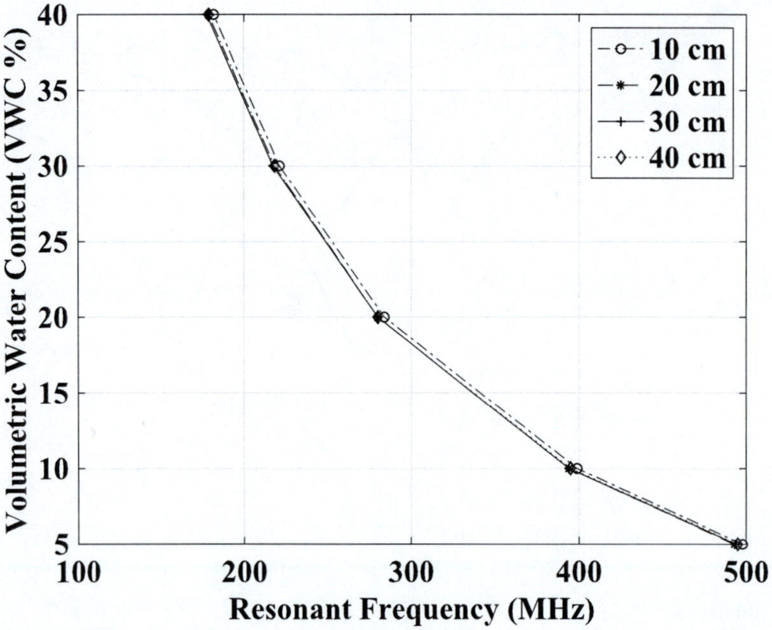

Fig. 7.7 Silty Clay Loam: change in soil moisture with resonant frequency at different depths

Figure 7.9 represents the graph for frequency with return loss. Return loss of a signal is defined as a proportion of signal waves rejected against accepted waves at the antenna. Hence, lower values of return loss are preferred. Experiments are performed under varying soil moisture to observe the effect of frequency along with the soil moisture content on return loss of a signal. Frequency is inversely proportional to VWC. As the soil moisture content decreases, the frequency increases. For example, frequency is maximum (approximately **150** MHz) at the minimum soil moisture value, i.e., VWC = **5** %, and the maximum value of frequency is **400** MHz at the highest VWC, i.e., **40** %. Return loss is maximum at lower VWC and higher frequency, and however, difference is not significant.

7.4.5 Sandy Clay

Figure 7.10 plots the resonant frequency with VWC values for different burial depths of **10** cm, **20** cm, **30** cm, and **40** cm. It can be seen that, for all burial depths, resonant frequency increases as VWC decreases. A significant increase is observed in frequency when VWC decreases from **20**–10%, i.e., approximately **90** MHz. The minimum change in frequency is observed between VWC range of **40** and 30%. Another important behavior that can be observed in the figure is that under burial

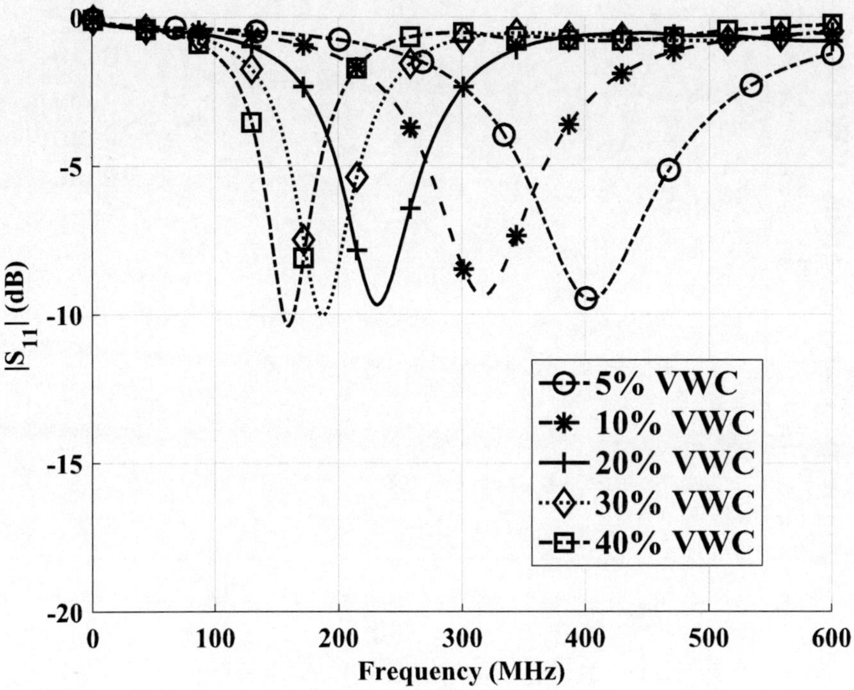

Fig. 7.8 Sandy Loam: resonant frequency with return loss for different soil moisture levels

depths of **20** cm, **30** cm, and **40** cm, the trend line remains approximately the same. However, at **10** cm the resonant frequency is a slightly higher as compared to other data points at the same VWC percentage. Therefore, it can be summarized that resonant frequency increases under low VWC and lower antenna's depth.

Figure 7.11 represents the graph for frequency with return loss. Return loss of a signal is defined as a proportion of signal waves rejected against accepted waves at the antenna. Hence, lower values of return loss are preferred. Experiments are performed under varying soil moisture to observe the effect of frequency along with the soil moisture content on return loss of a signal. Frequency is inversely proportional to VWC. As the soil moisture content decreases, frequency increases. For example, frequency is maximum (approximately **140** MHz) at the minimum soil moisture value, i.e., VWC = **5** %, and the maximum value of frequency is **400** MHz at the highest VWC, i.e., **40** %. Return loss is maximum at lower VWC and higher frequency, and however, the difference is not significant.

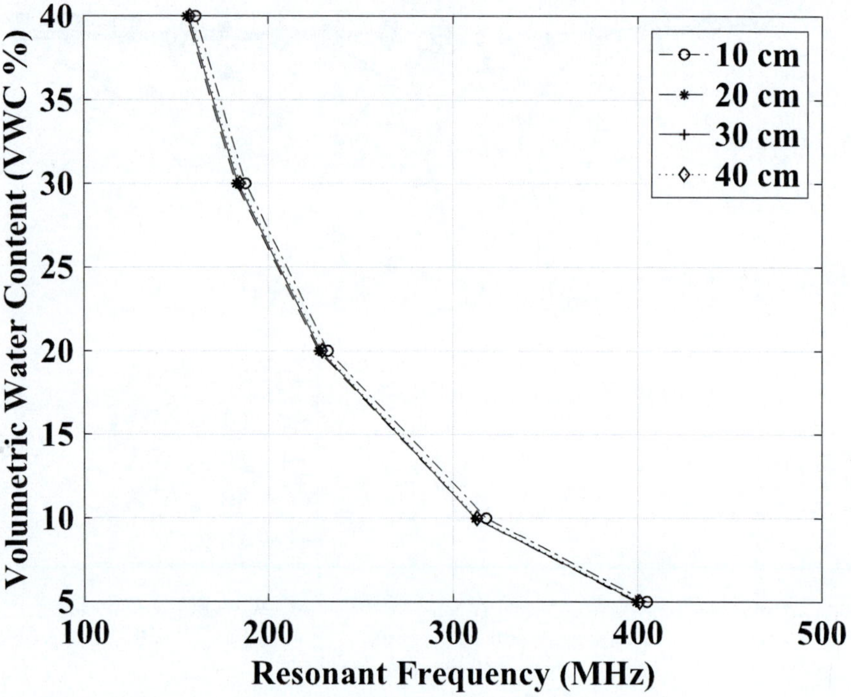

Fig. 7.9 Sandy Loam: change in soil moisture with resonant frequency at different depths

7.4.6 *Sandy Clay Loam*

Figure 7.12 plots resonant frequency with VWC values under the burial depth of **10** cm, **20** cm, **30** cm, and **40** cm. The overall trend between resonant frequency and the VWC remains the same for all burial depths, i.e., resonant frequency increases as the VWC decreases. The biggest change (approximately **80** MHz) in resonant frequency can be observed when VWC decreases from **20** % to **10** %. During this interval resonant frequency goes from **230** to 310MHz. A similar trend is observed when VWC further decreases from **10** % to **5** % and resonant frequency increases from approximately **310**–390MHz. For other ranges of VWC, i.e., **40**–30% and **30**–20%, the resonant frequency increases gradually. This shows that for lower VWC in soil, the resonant frequency increases significantly and it increases slowly as the soil's VWC increases.

Figure 7.13 plots the return loss with frequency under different VWC values, i.e., **5**–40%. It is important to note that higher values of return loss are more preferable. The experiment shows that the overall trend between frequency and the return loss remains the same for all VWC values, i.e., the return loss is higher at higher values of frequency. The lowest value of return loss is approximately **−7.5** dB which is observed at **40** % VWC and **150** MHz. Return loss is the highest (**−6** dB) at **5** %

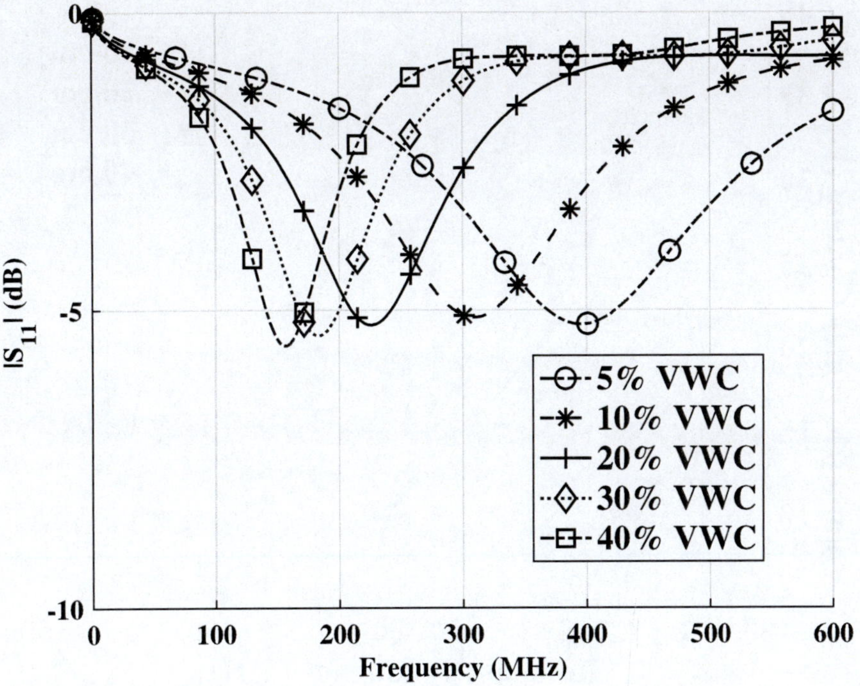

Fig. 7.10 Sandy Clay: resonant frequency with return loss for different soil moisture levels

VWC and **380** MHz. This shows that signal strength will increase at lower VWC values and higher frequencies and it will decrease at higher VWC values and low frequencies.

7.4.7 Loamy Sand

Figure 7.14 plots resonant frequency with VWC values for different burial depths of **10** cm, **20** cm, **30** cm, and **40** cm. It can be seen that, for all burial depths, resonant frequency increases as VWC decreases. A significant decrease is observed in frequency when VWC decreases from **20**–10%, i.e., approximately **90** MHz. The minimum change in frequency is observed between VWC range of **40** and 30%. This behavior shows that as the water content in the soil increases, values of resonant frequency tends to decrease and vice versa.

Figure 7.15 plots the return loss with frequency under different VWC values, i.e., **5**–40%. It is important to note that higher values of return loss are more preferable. The experiment shows that the overall trend between frequency and the return loss remains same for all values of VWC, i.e., return loss attains higher values as the frequency increases. The lowest value of return loss is observed at **40** % VWC

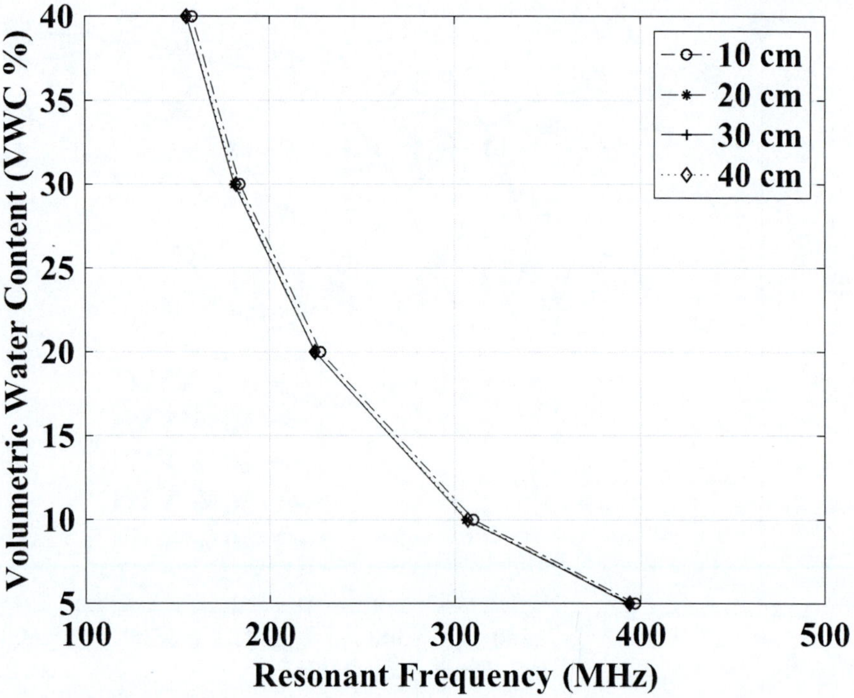

Fig. 7.11 Sandy Clay: change in soil moisture with resonant frequency at different depths

and **150** MHz. Return loss is highest at **5** % VWC and **380** MHz. This shows that signal strength will increase at lower VWC values and higher frequencies and it will decrease at higher VWC values and low frequencies.

7.4.8 Loam

Figure 7.16 plots resonant frequency with VWC values for different burial depths of **10** cm, **20** cm, **30** cm, and **40** cm. It can be seen that, for all burial depths, resonant frequency increases as VWC decreases. A significant increase is observed in frequency when VWC decreases from **20**–10%, i.e., approximately **90** MHz. The minimum change in frequency is observed between VWC range of **40** and 30%. Another important behavior that can be observed in the figure is that under burial depths of **20** cm, **30** cm, and **40** cm, the trend line remains approximately the same. However, at **10** cm the resonant frequency is a little bit higher as compared to other data points at the same VWC percentage. Therefore, it can be summarized that resonant frequency increases under low VWC and lower antenna's depth.

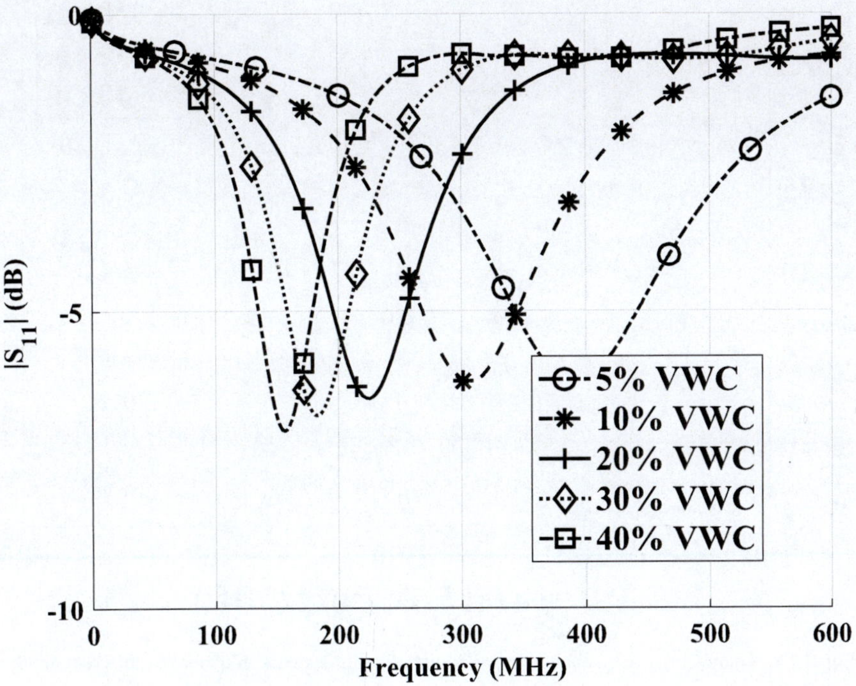

Fig. 7.12 Sandy Clay Loam: resonant frequency with return loss for different soil moisture levels

Figure 7.17 represents the graph for frequency with return loss. Return loss of a signal is defined as a proportion of signal waves rejected against accepted waves at the antenna. Hence, lower values of return loss are preferred. Experiments are performed under varying soil moisture to observe the effect of frequency along with the soil moisture content on return loss of a signal. Frequency is inversely proportional to VWC. As the soil moisture content decreases, frequency increases. For example, frequency is maximum (approximately **160** MHz) at the minimum soil moisture value, i.e., VWC = **5** %, and the maximum value of frequency is **420** MHz at the highest VWC, i.e., **40** %. Return loss is maximum at lower VWC and higher frequency, and however, the difference is not significant.

7.4.9 Clay

Figure 7.18 shows the graph plotted from experiments performed under varying VWC percentages of soil. Return loss was observed for different frequencies and changing VWCs. In simple terms, return loss represents the signal strength. Therefore, higher values of return loss means the signal is powerful. To that end,

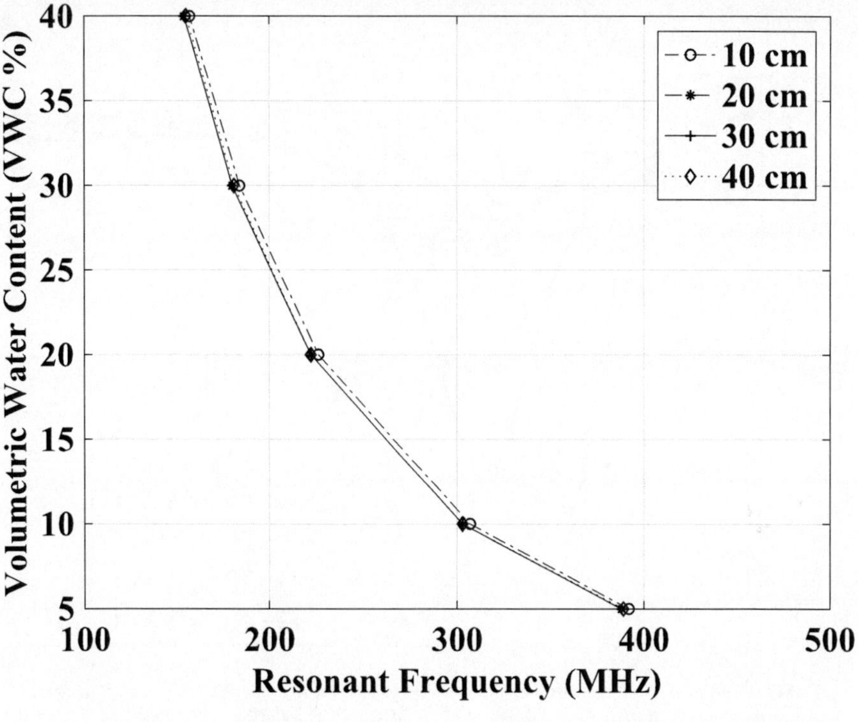

Fig. 7.13 Sandy Clay Loam: change in soil moisture with resonant frequency at different depths

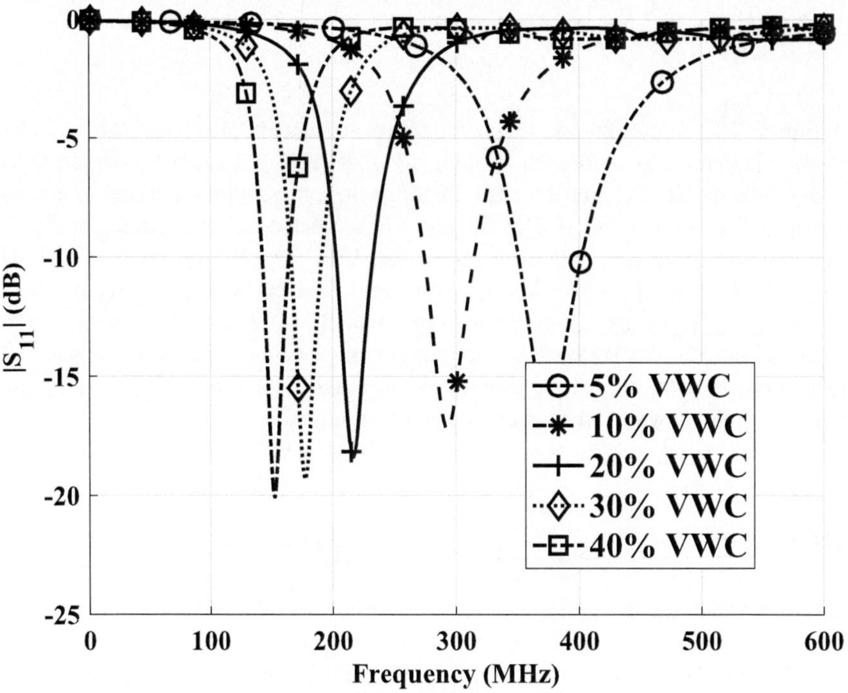

Fig. 7.14 Loamy Sand: resonant frequency with return loss for different soil moisture levels

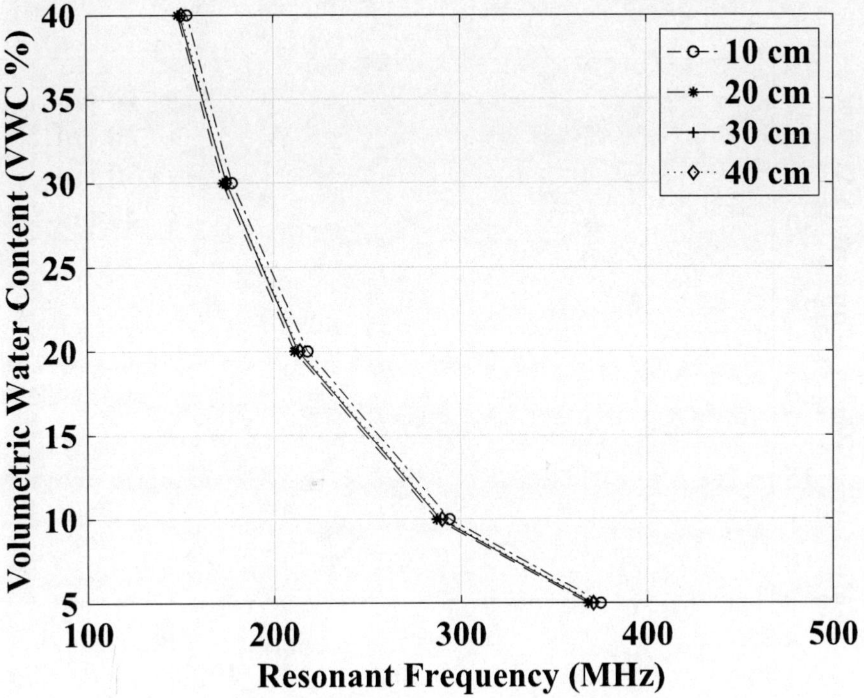

Fig. 7.15 Loamy Sand: change in soil moisture with resonant frequency at different depths

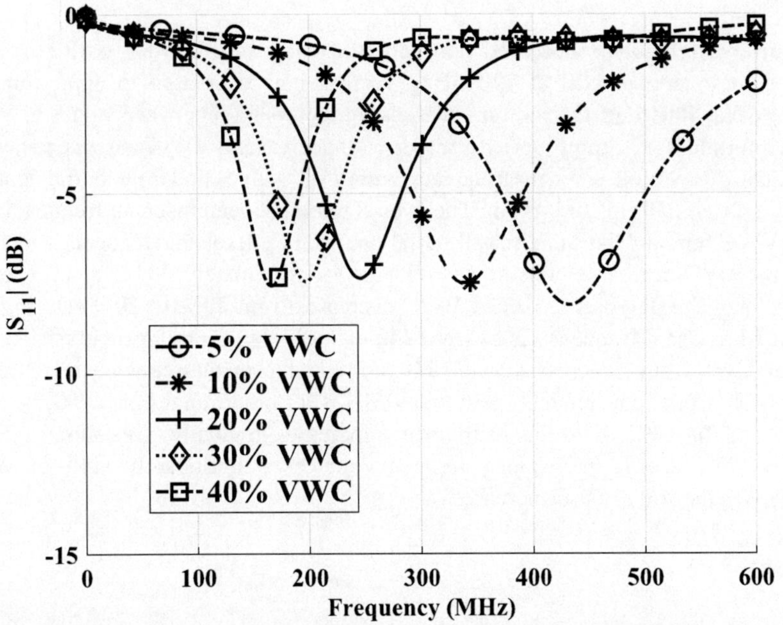

Fig. 7.16 Loam: resonant frequency with return loss for different soil moisture levels

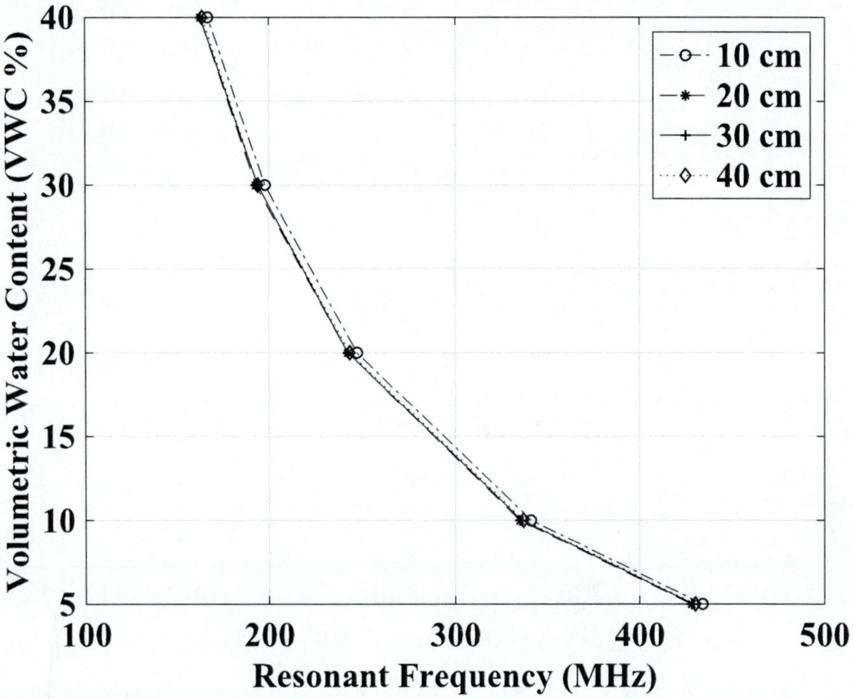

Fig. 7.17 Loam: change in soil moisture with resonant frequency at different depths

an inverse relation is observed between frequency and return loss. Return loss is approximately **−4** dB at **120** MHz, whereas it is decreased to approximately **−4.9** dB at **400** MHz. On contrary, there is a direct relation between values of VWC and return loss, i.e., return loss decreases with decrease in VWC and vice versa.

Figure 7.19 plots resonant frequency with VWC values under the burial depth of **10** cm, **20** cm, **30** cm, and **40** cm. The overall trend between resonant frequency and the VWC remains the same for all burial depths, i.e., resonant frequency increases as the VWC decreases. The biggest change (approximately **90** MHz) in resonant frequency can be observed when VWC decreases from **20** % to **10** %. During this interval resonant frequency goes from **240** to 330MHz. A similar trend is observed when VWC further decreases from **10** % to **5** % and resonant frequency increases by **100** MHz from approximately **330** to 430MHz. For other ranges of VWC, i.e., **40**–30% and **30**–20%, the resonant frequency increases gradually. This shows that for lower VWC in soil, the resonant frequency increases significantly and it increases slowly as the soil's VWC increases.

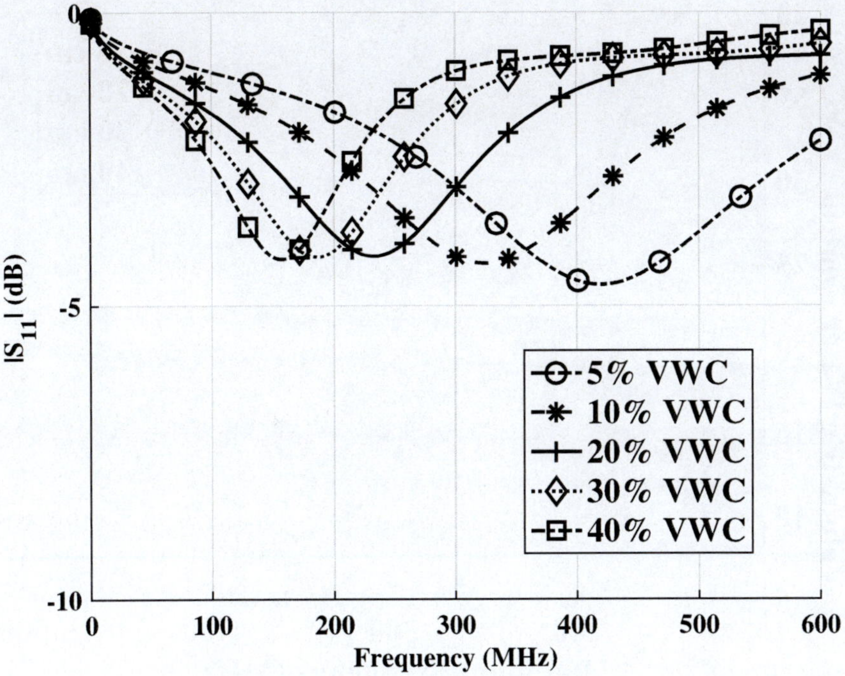

Fig. 7.18 Clay: resonant frequency with return loss for different soil moisture levels

7.4.10 Clay Loam

Figure 7.20 shows the graph plotted from experiments performed under varying VWC percentages of soil. Return loss was observed for different frequencies and changing VWCs. In simple terms, return loss represents the signal strength. Therefore, higher values of return loss means the signal is powerful. To that end, an inverse relation is observed between frequency and return loss. Return loss is approximately **−5.5** dB at **160** MHz, whereas it is decreased to approximately **−7** dB at **420** MHz. On contrary, there is a direct relation between the values of VWC and return loss, i.e., return loss decreases with decrease in VWC and vice versa. Moreover, Fig. 7.21 shows similar results for clay loam soil moisture sensing.

7.5 Discussion and Conclusions

In this chapter, considering the smart agriculture communication system, a performance analysis of dipole antenna return loss is done for the purpose of using antennas as sensors in smart agriculture. To that end, experiments are performed

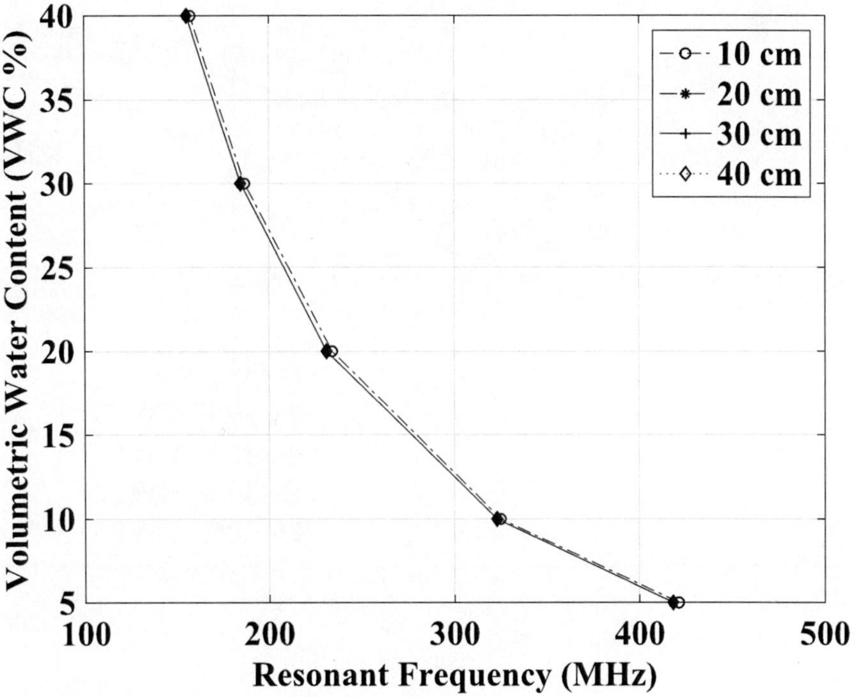

Fig. 7.19 Clay: change in soil moisture with resonant frequency at different depths

under varying experimental parameters. The parameters considered are soil type, soil moisture, and operational frequencies. The model is evaluated for resonant frequencies under varying values of all these parameters. Other important aspect of IOUT system is the relation between the resonant frequency and over-the-air (OTA) frequencies. Moreover, it is also important to perform a comparative analysis of the proposed model with permittivity-based antenna without using burial depths. All this is done to understand and propose a solid IOUT system design. The discussion is concluded with the accomplishment and results of the experiments and the research problems that can be considered as a future work to further improve the process of measuring soil moisture using radio wave propagation.

Experiments are performed in different soils for return loss and resonant frequency. For both types of soil, soil moisture ranges from **5** % to **40** %. The experiment is performed at the frequencies of **433** MHz and **915** MHz. For sandy soil, it can be concluded from figures that with an increase in soil moisture, resonant frequency decreases by **59** %. Similarly, for silt loam soil, resonant frequency decreases by **62** %. f_{rs} represents the resonant frequency and f_{ro} represents the OTA resonant frequency. It is shown that the ratio decreases with the increase in soil moisture value. This decrease is because of the decrease in resonant frequency.

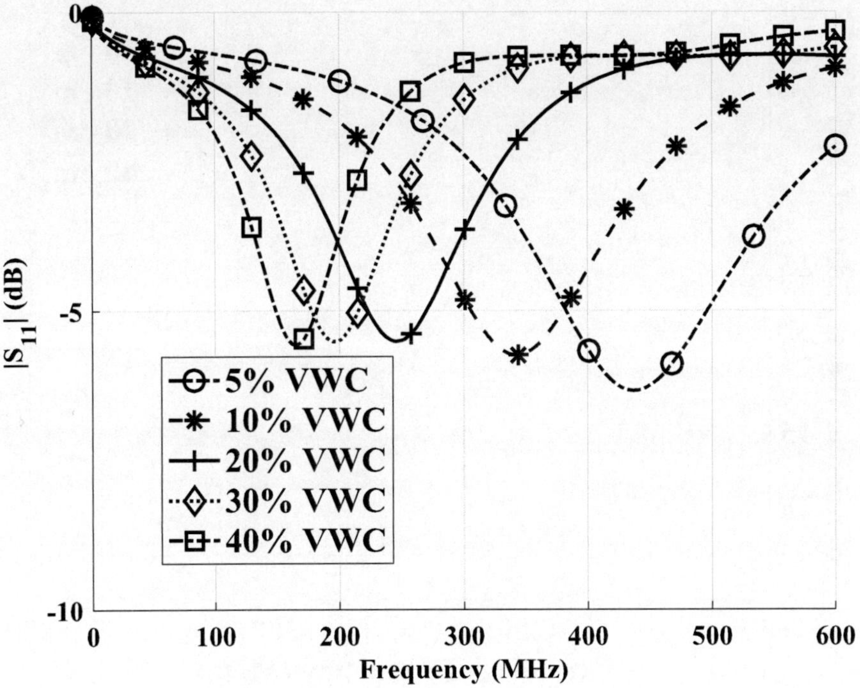

Fig. 7.20 Clay Loam: resonant frequency with return loss for different soil moisture levels

Moreover, ration is different for different values of burial depth at both frequencies of **433** MHz and **915** MHz.

Three measurement parameters, i.e., antenna T-R separation distance, frequency, and height, were selected based on theoretical discussion. Although any frequency greater than **30** MHz was found to be optimum, however, a **170** MHz was selected to satisfy the requirement of equipment and antenna size for that frequency. After experimentation, it was found that the optimum value for antenna separation is around 20 wavelengths. Hence, a separation distance of 17.3 wavelength, i.e., **100** ft, for a frequency of **170** MHz. It was required that the space wave should be negligible as compared to the surface wave. To satisfy this requirement, antenna with the height of $\frac{1}{8}$ wavelength was selected. After experiments, it was concluded that antenna impedance is inversely proportional to the its height. Various other variables, apart from soil moisture, were also investigated to see their effect on radio wave propagation. Among those variables, antenna height was found to be very important. Error occurred due to reflection if the objects were moved within **50** ft of the antennas. It was concluded, from experiments, that **100** ft is the optimal distance. Since the temperature had negligible effect on the equipment, it was not considered as a measurement parameter. Vegetation affects the signal and soil moisture to great extent, and therefore, it was removed from experiments.

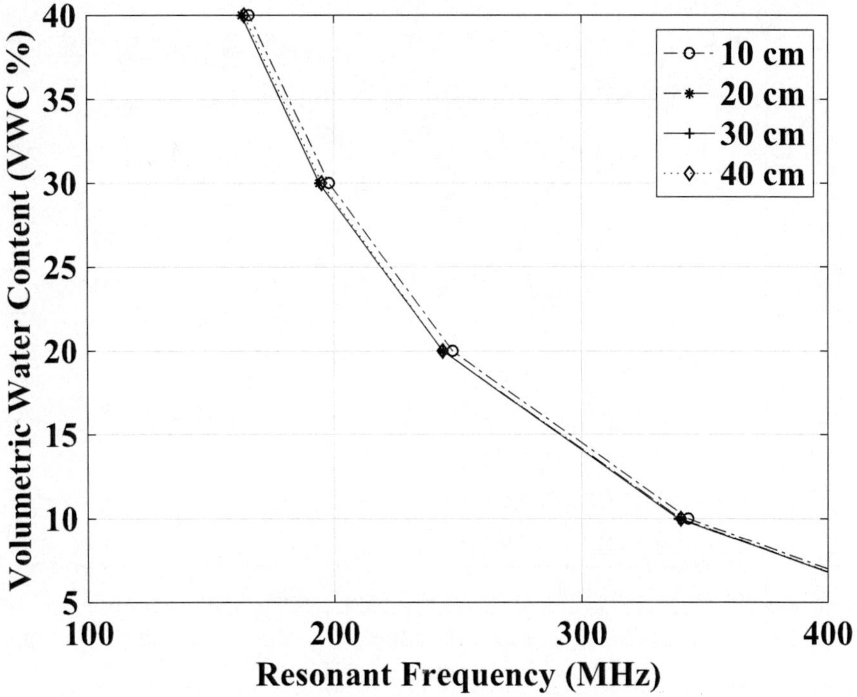

Fig. 7.21 Clay Loam: change in soil moisture with resonant frequency at different depths

Experiments performed to compare return loss and soil moisture gave promising results. Most of the measurements show that increase in soil moisture results in increase in permittivity. Under **1.67** in of precipitation, due to heavy rainstorm, signal strength doubled. Therefore, a positive correlation was observed between permittivity and freshly deposited moisture. However, it may cause the problem if moisture starts to penetrate deeply after some time. This can be seen from the outlier in some of the figures presented in this chapter.

Some questions, i.e., vegetation, and organic matter in soil layers due to soil moisture are left for the future work. The soil with non-homogeneous moisture may face problem due to these issues. Vegetation has strong effect on signal strength and it is important to differentiate its effect from the soil moisture. Dielectric constant is a major characteristics which is affected by soil moisture and may affect the signal attenuation. From these experiments, we have shown that the measurement of soil moisture can be done through underground antennas.

References

1. T.S. Bird, Definition and Misuse of Return Loss [Report of the Transactions Editor-in-Chief]. IEEE Antennas Propag. Mag. **51**, 166–167 (2009). https://doi.org/10.1109/MAP.2009.5162049
2. M. Dobson, et.al., Microwave Dielectric Behavior of Wet Soil—Part II: Dielectric Mixing Models. IEEE Trans. Geosci. Remote Sens. **GE-23**, 35–46 (1985). https://doi.org/10.1109/TGRS.1985.289498
3. X. Dong, M.C. Vuran, Return loss analysis of antennas in wireless underground sensor networks. submitted.
4. X. Dong, M.C. Vuran, Impacts of soil moisture on cognitive radio underground networks, in *Proceedings of the IEEE BlackSeaCom* (2013)
5. B. Josephson, A. Blomquist, The influence of moisture in the ground, temperature and terrain on ground wave propagation in the VHF-band. IRE Trans. Anntenas Propag. **6**, 169–172 (1958). https://doi.org/10.1109/TAP.1958.1144574
6. R.W.P. King, M. Owens, T.T. Wu, *Lateral Electromagnetic Waves* (Springer-Verlag, Berlin, 1992)
7. A. Konda, A. Rau, M.A. Stoller, J.M. Taylor, A. Salam, G.A. Pribil, C. Argyropoulos, S.A. Morin, Soft microreactors for the deposition of conductive metallic traces on planar, embossed, and curved surfaces. Adv. Funct. Mater. **28**, 1803020 (2018). https://doi.org/10.1002/adfm.201803020
8. U. Raza, A. Salam, Wireless underground communications in sewer and stormwater overflow monitoring: radio waves through soil and asphalt medium. Information **11**(2), 98 (2020)
9. U. Raza, A. Salam, Zenneck waves in decision agriculture: an empirical verification and application in EM-Based underground wireless power transfer. Smart Cities **3**, 308–340 (2020). https://doi.org/10.3390/smartcities3020017
10. U. Raza, A. Salam, On-Site and external power transfer and energy harvesting in underground wireless. Electronics **9**(4), 681 (2020)
11. U. Raza, A. Salam, A survey on subsurface signal propagation. Smart Cities **3**, 1513–1561 (2020). https://doi.org/10.3390/smartcities3040072
12. A. Salam, Underground soil sensing using subsurface radio wave propagation, in *5th Global Workshop on Proximal Soil Sensing* (2019)
13. A. Salam, A comparison of path loss variations in soil using planar and dipole antennas, in *2019 IEEE International Symposium on Antennas and Propagation* (IEEE, New York, 2019)
14. A. Salam, A path loss model for through the soil wireless communications in digital agriculture, in *2019 IEEE International Symposium on Antennas and Propagation* (IEEE, New York, 2019)
15. A. Salam, Underground environment aware MIMO design using transmit and receive beamforming in internet of underground things, in *2019 International Conference on Internet of Things (ICIOT 2019)* (2019)
16. A. Salam, An underground radio wave propagation prediction model for digital agriculture. Information **10**(4), 147 (2019). https://doi.org/10.3390/info10040147
17. A. Salam, Subsurface MIMO: a beamforming design in internet of underground things for digital agriculture applications. J. Sens. Actuator Netw. **8**(3), 41 (2019). https://doi.org/10.3390/jsan8030041
18. A. Salam, Design of subsurface phased array antennas for digital agriculture applications, in *Proceedings of the 2019 IEEE International Symposium on Phased Array Systems and Technology (IEEE Array 2019)* (2019)
19. A. Salam, *Internet of Things for Sustainable Community Development*, 1 edn. (Springer Nature, Berlin, 2020). https://doi.org/10.1007/978-3-030-35291-2
20. A. Salam, Pulses in the sand: long range and high data rate communication techniques for next generation wireless underground networks, in *ETD collection for University of Nebraska—Lincoln* (2018)

21. A. Salam, Sensor-free underground soil sensing, in *ASA, CSSA and SSSA International Annual Meetings (2019)*. ASA-CSSA-SSSA (2019)
22. A. Salam, Internet of Things for sustainable community development: introduction and overview, in *Internet of Things for Sustainable Community Development: Wireless Communications, Sensing, and Systems* (Springer International Publishing, Cham, 2020), pp. 1–31. https://doi.org/10.1007/978-3-030-35291-2_1
23. A. Salam, Internet of Things for environmental sustainability and climate change, in *Internet of Things for Sustainable Community Development: Wireless Communications, Sensing, and Systems* (Springer International Publishing, Cham, 2020), pp. 33–69. https://doi.org/10.1007/978-3-030-35291-2_2
24. A. Salam, Internet of Things in agricultural innovation and security, in *Internet of Things for Sustainable Community Development: Wireless Communications, Sensing, and Systems* (Springer International Publishing, Cham, 2020), pp. 71–112. https://doi.org/10.1007/978-3-030-35291-2_3
25. A. Salam, Internet of Things for water sustainability, in *Internet of Things for Sustainable Community Development: Wireless Communications, Sensing, and Systems* (Springer International Publishing, Cham, 2020), pp. 113–145. https://doi.org/10.1007/978-3-030-35291-2_4
26. A. Salam, Internet of Things for sustainable forestry, in *Internet of Things for Sustainable Community Development: Wireless Communications, Sensing, and Systems* (Springer International Publishing, Cham, 2020), pp. 147–181. https://doi.org/10.1007/978-3-030-35291-2_5
27. A. Salam, Internet of Things in sustainable energy systems, in *Internet of Things for Sustainable Community Development: Wireless Communications, Sensing, and Systems* (Springer International Publishing, Cham, 2020), pp. 183–216. https://doi.org/10.1007/978-3-030-35291-2_6
28. A. Salam, Internet of Things for sustainable human health, in *Internet of Things for Sustainable Community Development: Wireless Communications, Sensing, and Systems* (Springer International Publishing, Cham, 2020), pp. 217–242. https://doi.org/10.1007/978-3-030-35291-2_7
29. A. Salam, Internet of Things for sustainable mining, in *Internet of Things for Sustainable Community Development: Wireless Communications, Sensing, and Systems* (Springer International Publishing, Cham, 2020), pp. 243–271. https://doi.org/10.1007/978-3-030-35291-2_8
30. A. Salam, Internet of Things in water management and treatment, in *Internet of Things for Sustainable Community Development: Wireless Communications, Sensing, and Systems* (Springer International Publishing, Cham, 2020), pp. 273–298. https://doi.org/10.1007/978-3-030-35291-2_9
31. A. Salam, Internet of Things for sustainability: perspectives in privacy, cybersecurity, and future trends, in *Internet of Things for Sustainable Community Development: Wireless Communications, Sensing, and Systems* (Springer International Publishing, Cham, 2020), pp. 299–327. https://doi.org/10.1007/978-3-030-35291-2_10
32. A. Salam, U. Karabiyik, A cooperative overlay approach at the physical layer of cognitive radio for digital agriculture, in *Third International Balkan Conference on Communications and Networking 2019 (BalkanCom'19)* (2019)
33. A. Salam, S. Shah, Internet of Things in smart agriculture: enabling technologies, in *2019 IEEE 5th World Forum on Internet of Things (WF-IoT) (WF-IoT 2019)* (2019)
34. A. Salam, M.C. Vuran, Impacts of soil type and moisture on the capacity of multi-carrier modulation in internet of underground things, in *Proceeding of the 25th ICCCN 2016* (2016)
35. A. Salam, M.C. Vuran, Smart underground antenna arrays: a soil moisture adaptive beamforming approach, in *Proceeding IEEE INFOCOM 2017* (2017)
36. A. Salam, M.C. Vuran, Wireless underground channel diversity reception with multiple antennas for internet of underground things, in *Proceeding of the IEEE ICC 2017* (2017)
37. A. Salam, M.C. Vuran, EM-Based wireless underground sensor networks, in *Underground Sensing*, ed. by S. Pamukcu, L. Cheng (Academic Press, New York, 2018), pp. 247–285. https://doi.org/10.1016/B978-0-12-803139-1.00005-9
38. A. Salam, U. Raza, On Burial depth of underground antenna in soil horizons for decision agriculture, in 2020 International Conference on Internet of Things (ICIOT-2020) (2020)

39. A. Salam, U. Raza, *Signals in the Soil*, 1 edn. (Springer Nature, New York, 2020). https://doi.org/10.1007/978-3-030-50861-6
40. A. Salam, U. Raza, Signals in the soil: an introduction to wireless underground communications, in *Signals in the Soil: Developments in Internet of Underground Things* (Springer International Publishing, Cham, 2020), pp. 3–38. https://doi.org/10.1007/978-3-030-50861-6_1
41. A. Salam, U. Raza, Electromagnetic characteristics of the soil, in *Signals in the Soil: Developments in Internet of Underground Things* (Springer International Publishing, Cham, 2020), pp. 39–59. https://doi.org/10.1007/978-3-030-50861-6_2
42. A. Salam, U. Raza, Wireless underground channel modeling, in *Signals in the Soil: Developments in Internet of Underground Things* (Springer International Publishing, Cham, 2020), pp. 61–121. https://doi.org/10.1007/978-3-030-50861-6_3
43. A. Salam, U. Raza, Modulation schemes and connectivity in wireless underground channel, in *Signals in the Soil: Developments in Internet of Underground Things* (Springer International Publishing, Cham, 2020), pp. 125–166. https://doi.org/10.1007/978-3-030-50861-6_4
44. A. Salam, U. Raza, Underground wireless channel bandwidth and capacity, in *Signals in the Soil: Developments in Internet of Underground Things* (Springer International Publishing, Cham, 2020), pp. 167–188. https://doi.org/10.1007/978-3-030-50861-6_5
45. A. Salam, U. Raza, Signals in the soil: underground antennas, in *Signals in the Soil: Developments in Internet of Underground Things* (Springer International Publishing, Cham, 2020), pp. 189–215. https://doi.org/10.1007/978-3-030-50861-6_6
46. A. Salam, U. Raza, Underground phased arrays and beamforming applications, in *Signals in the Soil: Developments in Internet of Underground Things* (Springer International Publishing, Cham, 2020), pp. 217–248. https://doi.org/10.1007/978-3-030-50861-6_7
47. A. Salam, U. Raza, Signals in the soil: subsurface sensing, in *Signals in the Soil: Developments in Internet of Underground Things* (Springer International Publishing, Cham, 2020), pp. 251–297. https://doi.org/10.1007/978-3-030-50861-6_8
48. A. Salam, U. Raza, Soil moisture and permittivity estimation, in *Signals in the Soil: Developments in Internet of Underground Things* (Springer International Publishing, Cham, 2020), pp. 299–317. https://doi.org/10.1007/978-3-030-50861-6_9
49. A. Salam, U. Raza, Current advances in internet of underground things, in *Signals in the Soil: Developments in Internet of Underground Things* (Springer International Publishing, Cham, 2020), pp. 321–356. https://doi.org/10.1007/978-3-030-50861-6_10
50. A. Salam, U. Raza, Decision agriculture, in *Signals in the Soil: Developments in Internet of Underground Things* (Springer International Publishing, Cham, 2020), pp. 357–378. https://doi.org/10.1007/978-3-030-50861-6_11
51. A. Salam, U. Raza, Autonomous irrigation management in decision agriculture. in *Signals in the Soil: Developments in Internet of Underground Things* (Springer International Publishing, Cham, 2020), pp. 379–398. https://doi.org/10.1007/978-3-030-50861-6_12
52. A. Salam, U. Raza, Variable rate applications in decision agriculture, in *Signals in the Soil: Developments in Internet of Underground Things* (Springer International Publishing, Cham, 2020), pp. 399–423. https://doi.org/10.1007/978-3-030-50861-6_13
53. A. Salam, M.C. Vuran, S. Irmak, Pulses in the sand: impulse response analysis of wireless underground channel, in *The 35th Annual IEEE International Conference on Computer Communications (INFOCOM 2016)* (2016)
54. A. Salam, M.C. Vuran, S. Irmak, Towards internet of underground things in smart lighting: a statistical model of wireless underground channel, in *Proceedings of the 14th IEEE International Conference on Networking, Sensing and Control (IEEE ICNSC)* (2017)
55. A. Salam, M.C. Vuran, S. Irmak, Di-Sense: in situ real-time permittivity estimation and soil moisture sensing using wireless underground communications. Comput. Netw. **151**, 31–41 (2019). https://doi.org/10.1016/j.comnet.2019.01.001
56. A. Salam, M.C. Vuran, X. Dong, C. Argyropoulos, S. Irmak, A theoretical model of underground dipole antennas for communications in internet of underground things. IEEE Trans. Antennas Propag. **67**(6), 3996–4009 (2019)

57. A. Salam, A.D. Hoang, A. Meghna, D.R. Martin, G. Guzman, Y.H. Yoon, J. Carlson, J. Kramer, K. Yansi, M. Kelly, et al., The Future of Emerging IoT Paradigms: Architectures and Technologies (2019)
58. A. Salam, M.C Vuran, S. Irmak, A statistical impulse response model based on empirical characterization of wireless underground channel. IEEE Trans. Wirel. Commun. **19**, 5966–5981 (2020)
59. S. Temel, M.C. Vuran, M.M. Lunar, Z. Zhao, A. Salam, R.K. Faller, C. Stolle, Vehicle-to-barrier communication during real-world vehicle crash tests. Comput. Commun. **127**, 172–186 (2018). https://doi.org/10.1016/j.comcom.2018.05.009
60. G.C. Topp, J.L. Davis, A.P. Annan, Electromagnetic determination of soil water content: measurements in coaxial transmission lines. Water Resour. Res. **16**, 574–582 (1980). https://doi.org/10.1029/WR016i003p00574.
61. M.C. Vuran, A. Salam, R. Wong, S. Irmak, Internet of Underground Things: Sensing and Communications on the Field for Precision Agriculture, in *2018 IEEE 4th World Forum on Internet of Things (WF-IoT) (WF-IoT 2018)* (2018)
62. M.C. Vuran, A. Salam, R. Wong, S. Irmak, Internet of Underground Things in Precision Agriculture: Architecture and Technology Aspects, in *Ad Hoc Networks* (2018). https://doi.org/10.1016/j.adhoc.2018.07.017

Chapter 8
Smart Sewer Experimental Results

8.1 Introduction

IOUT is a new and evolving paradigm. Many of the potential applications in IOUT need theoretical explanation and model development of the underground channel. These applications also require detailed empirical evaluations of the wireless underground communication channel. Wireless communication is affected by the properties of the medium. Wireless underground communication is carried out through soil medium and is very sensitive to the physical properties of soil and soil moisture. and most of the communication is carried out through soil medium. Therefore, conducting wireless channel experiment outdoor can be a very daunting task because of the changing weather condition and soil properties. One solution to cope with these challenges is to develop indoor testbed. Indoor testbeds are used for conducting the experiments on the wireless underground channel characterization. There has been an extensive effort done for the channel characterization, and measurements have been taken over the span of 3 years. Different types of the soils (silty clay loam, silt loam, and sandy soils) are used to empirically evaluate the underground antenna performance and channel transfer function. The experiments have shown that indoor testbeds are a fast and efficient way of underground communication characterization. Furthermore, multiple types of soils can be used for experimentation using indoor testbeds. Underground communication and antenna performance are affected by the different properties of soil, e.g., moisture, texture, and burial depth. The experiments were done to perform the power delay profile analysis of the UG channel. The analysis shows that UG communication uses different methods for short- (through-the-earth, i.e., direct wave) and long-distance (up-over-down; lateral wave) communications. These experiments show the importance of considering the soil physical properties for analyzing UG communication. Consequently, an improved UG communication method can facilitate the implementation of efficient and practical IOUT systems [2, 3, 5, 27].

A. Salam, *Internet of Things in Smart Sewer and Drainage Systems*,
https://doi.org/10.1007/978-3-031-48109-3_8

8.2 Experimental Setup

8.2.1 The Indoor Testbed

Outdoor experimentation in WUSN is not an easy task [28–32]. Outdoor settings face challenges of extreme weather and temperature conditions. It is very important to get the timely results of experiments, and getting the different soil moisture level in short span of time is difficult in outdoor settings. Furthermore, it lacks dynamic soil moisture control, difficult to get changing soil types and deployment of equipment is also cumbersome. An indoor testbed can overcome these challenges. Figure 8.1 shows an indoor testbed. This indoor testbed is developed in a greenhouse setting. It is a wooden box (100 in × 36 in × 48 in) with 90 ft^3 of soil in it (see Fig. 8.1a). It is equipped with a drainage system at the bottom and has waterproof sides. The sides are water proofed using water proof tarp to prevent leaks from the side. The box has a 3 in layer of gravel underneath to allow free water drainage (see Fig. 8.1b). Figure 8.1c shows the box with soil in it. It uses Watermark sensors to monitor the soil moisture level. A total of eight sensors are installed on the sides of the box. These sensors are placed at the depths of 10 cm, 20 cm, 30 cm, and 40 cm. Two Watermark dataloggers are connected to these sensors. For each installed antenna, a tamper tool is used to pack soil every 30 cm. It is done to achieve the bulk density[1] to simulate the real-world scenario. A total of 12 antennas, divided

Fig. 8.1 Field testbed development in the silty loam soil: (**a**) testbed layout, (**b**) antenna placement, (**c**) outlook after antenna installation, (**d**) antenna cables out of soil at different depths, and (**e**) USRPs and datalogger for soil moisture measurements

[1] Ratio of dry soil weight to the volume of soil (also including the volume of pores in particles).

Table 8.1 Particle size distribution and classification of testbed soils

Textural class	%Sand	%Silt	%Clay
Sandy soil	86	11	3
Silt loam	33	51	16
Silty clay loam	13	55	32

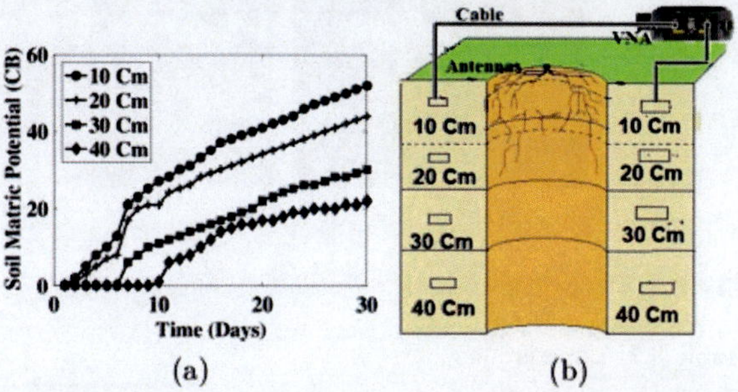

Fig. 8.2 (**a**) Soil moisture (expressed as soil matric potential; greater matric potential values indicate lower soil moisture and zero matric potential represents near-saturation condition) with time in silt loam testbed, (**b**) experiment layout

into sets of three, are installed (see Fig. 8.1d at increasing depths of 10 cm, 20 cm, 30 cm, and 40 cm. These sets are separated at 50 cm from each other. Figure 8.1e shows the final testbed [33–36].

Two different soil types were used for the experimentation: silt loam and sandy soil. Table 8.1 lists the soil type used and distribution of the particle size within the soil type. Experiments were performed to investigate the effect of soil texture on underground communications. To that end, soil having sand content from 13% to 86% and clay content from 3% to 32% was chosen. For experimentation, a nearly saturated soil is used as an input to achieve the maximum volumetric water content (VWC) level. Afterward, results are gathered as the soil water potential decreases from the saturated state to field capacity[2] and then finally to wilting point.[3] Figure 8.2a shows the moisture level change in silt loam soil.

8.2.2 The Field Testbed

A field testbed consists of dipole antennas, buried at 20 cm depth, with silty clay loam soil (see Fig. 8.3a). The purpose of this testbed is to perform underground-to-aboveground (UG2AG) experiments. Field testbed is also used to compare the results from the indoor testbed. For UG2AG experiments, an adjustable height pole is used. The experiments are conducted with the radii of 2 m, 4 m, 5.5 m, and 7 m.

---

[2] Water content in the soil after the drainage of excess water.

[3] The minimum level of water in the soil.

(a) (b)

Fig. 8.3 The field testbed

Table 8.2 Underground channel measurement parameters

Parameter	Value
Start frequency	10 MHz
Stop frequency	4 GHz
Number of frequency points	401
Transmit power	5 dBm
Vector network analyzer	Agilent FieldFox

The distance of max 73 is used due to cable limitations of antenna used for vector network analyzer (VNA). Receiver angels of 0°, 30°, 45°, 60°, and 90° are used. The testbed is shown in Fig. 8.3.

8.2.3 UG Software-Defined Radio (SDR)

Another testbed is used for conducting experiments with underground software-defined radios. This testbed consists of a total of 16 dipole antennas, divided into four sets, in silt loam soil. For each set, the antenna burial distance is 50 cm, 2 m, and 4 m from the first antenna. In each set, antennas are buried at the depths of 10 cm, 20 cm, 30 cm, and 40 cm. These dipole antennas use the resonant frequency of 433 MHz for over-the-air communication. Figure 8.1 shows-step by-step development of the testbed (starting form Fig. 8.1a–e). Figure 8.1a shows the layout, Figs. 8.1b, c show the testbed after placement and installation of the antenna, and Fig. 8.1d shows the cable coming out of the antennas buried at different depths. Finally, Fig. 8.1e shows the testbed with USRPSs and datalogger used for measuring soil moisture (Table 8.2).

8.2.4 *Soil Moisture Logging*

AS discussed previously, soil moisture has an impact on the communication. Therefore, it is important to log the soil moisture data after each experiment for correct characterization of channel. Oven drying method can be used for determining soil moisture, and however, it requires the removal of soil from the testbed. Watermark sensors are because of their ability of logging soil moisture data with timestamp. Watermark sensors are fast, efficient, and less error prone. It also overcomes the challenge of oven drying method. If metallic object soil is within the vicinity of buried antennas, they can cause interference during the communication. Therefore, to avoid interference, sensors are deployed at the edges of the testbed [37–41].

8.3 Measurement Techniques and Experiment Description

These testbeds use Keysight Technologies N9923A, FeildFox, and VNA for taking measurements. The layout of the measurements is shown in Fig. 8.2b. In the coming section, measurements taken for the indoor testbed are explained in detail.

8.3.1 *Path Loss Measurements*

VNA uses a UG transmitter and receiver (T–R) pair for measuring the signal loss. It transmits a known signal and the comparison of received and incident signals is performed to calculate the loss. Path loss is defined as the ratio of power of a signal at the sender end P_t to power of the signal reached at the receiver end P_r. Path loss is calculated by the following equation:

$$PL = P_t - P_r = 10. \log 10(P_t/P_r), \tag{8.1}$$

where PL denotes the path loss of the system. It also includes the effect of transmitting antenna gain G_t and receiving antenna gain G_r. The frequency range used for the measurement of path loss is 10 Mhz to 4 GHz. A total of 401 discrete frequency points were taken at varying distance and depths. These measurements were used to study how physical properties of the soil impact the attenuation of UG channel.

8.3.2 Power Delay Profile (PDP) Measurements

Power delay profile (PDP) is a way to measure the signal intensity through a multipath channel. The intensity is measured as a time delay. Time delay is the difference of time taken by signal to travel through multiple paths. The multipath characteristic of a wireless underground channel is studied by channel sounding experiment. VNA accurately characterizes the UG channel by transmitting multiple sine waves at the transmitter end and measuring them at the receiver end. These sine waves are from low to high frequencies. Impulse response of UG channel is measured by VNA using one frequency at time. It measures the impulse response in frequency domain instead of time domain.

8.4 Measurement Campaigns

This section presents the measurements parameters for experiments on each type of the soil.

8.4.1 Sandy Soil Experiments

For a distance of 50 cm between transmitter and receiver, antenna was buried at the depths of 10 cm, 20 cm, 30 cm, and 40 cm in a sandy soil. For a soil moisture range of 0–250 CB, they are buried at a depth of 1 m.

8.4.2 Silty Clay Loam Experiments

For a distance of 50 cm between transmitter and receiver, antenna was buried at the depth of 20 cm in a silty clay loam soil. In field testbed, it is buried at 1 m depth for a soil moisture range of 0–50 CB.

8.4.3 Silt Loam Experiments

For a distance of 50 cm between transmitter and receiver, antenna was buried at the depths of 10 cm, 20 cm, 30 cm, and 40 cm in a silt loam soil. In indoor testbed, for a soil moisture range of 0–50 CB, they are buried at a depth of 1 m.

8.4.4 *Path Loss Experiments for Different Aboveground Receiver Antenna Angles*

8.4.5 *Planar Antenna Experiments*

Indoor testbed and two planar antennas are used for the planar antenna experiments. These experiments are conducted in silty clay loam and sandy soils. The return loss and path loss are measured without obstructions and with obstruction between these two antennas. For the former experiment, the antenna was kept 1 m apart and buried at the depth of 20 cm. For later, another antenna is placed in the middle (50 cm) of the two antennas acting as an obstruction and the depth for all the antennas were kept the same (20 cm). These experiments were also conducted with silty clay loam soil using the same empirical parameters, and return loss and path loss are compared for both types of the soil.

8.5 Empirical Results

8.5.1 *Channel Transfer Function Measurements*

This section explores the channel transfer function (S_{12}) for different types of soil using varying values for distances, soil moisture, and depth.

8.5.1.1 Impact of Burial Depth and Antenna Distance on Attenuation

Figure 8.4 plots the attenuation at different depth and distance in silt loam soil. Figure 8.4a, b shows the attenuation, when the antennas are 50 cm apart with varying frequency and depths (10 cm, 20 cm, 30 cm, and 40 cm). It can be observed in these figures that at 200 MHz path loss differs by 5 dB at 10 cm (where path loss = 40.37 dB) from 40 cm (where path loss = 45.26 dB). The difference further increases to 8 dB at 250 MHz. Figure 8.4c, d shows the attenuation, when the antennas are 1 m apart with frequency and depths (10 cm, 20 cm, 30 cm, and 40 cm).

At 200 MHz, path loss for the depth of 10 cm is 44.37 dB and that for 40 cm is 60.12 dB. The difference in path loss is 16 dB which is higher as compared to difference at 50 cm. The difference increases to 20 dB at the frequency of 250 MHz.

Results for similar experiments with the same depths and frequencies, but with different soil type (silt loam soil), are shown in Figs. 8.5 and 8.6. Figure 8.5 shows the results with transmitter and receiver 50 cm apart and Fig. 8.6 shows for the distance of 1 m. For both distances attenuation is the highest at the frequency of 400 MHz for all depths.

Another important impact to consider is that of distance between the antennas on attenuation. From Fig. 8.4a (50 cm) and Fig. 8.4c (1 m), it can be observed that

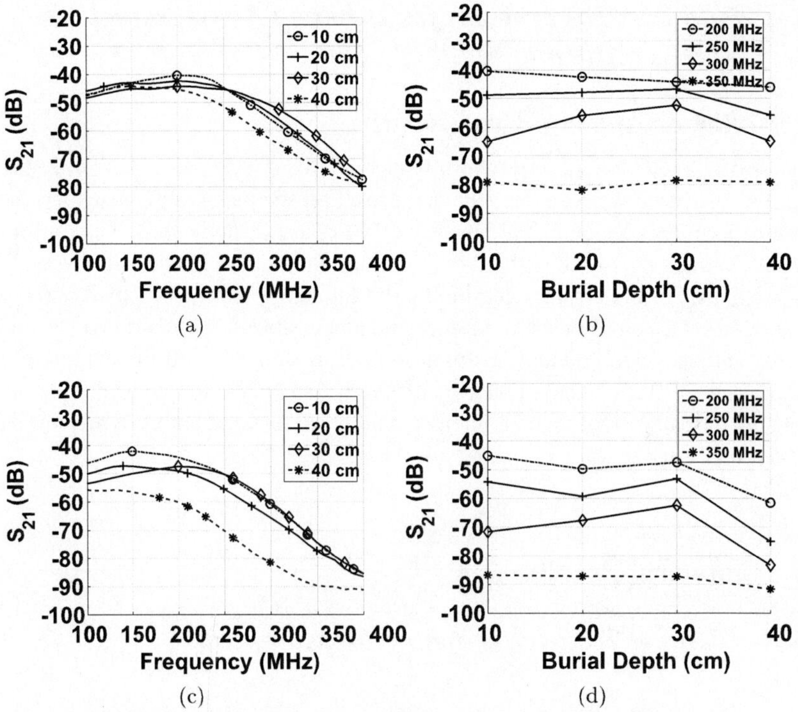

Fig. 8.4 Attenuation in silt loam soil at **50** cm distance: (**a**) with frequency, (**b**) at different depths, attenuation in silt loam soil at **1** distance, (**c**) with frequency, and (**d**) at different depths

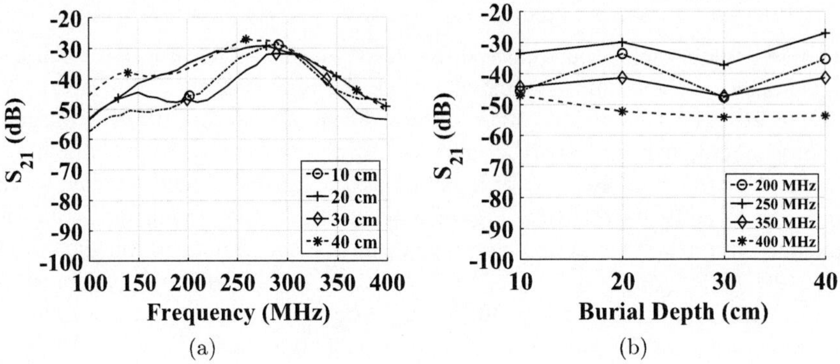

Fig. 8.5 Attenuation in sandy soil at **50** cm distance: (**a**) with frequency and (**b**) at different depths

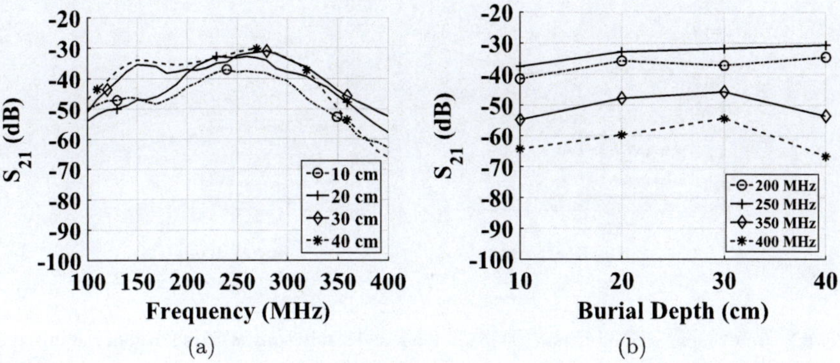

Fig. 8.6 Attenuation in sandy soil at **1** distance: (**a**) with frequency and (**b**) at different depths

path loss increases as the distance between antennas is increased from 50 cm to 1 m. For example, when the antenna distance increases from 50 cm to 1 m at 200 MHz, path loss is increased by 5 dB, 7 dB, 3 dB, and 15 dB at depths of 10 cm, 20 cm, 30 cm, and 40 cm, respectively. A similar trend can be observed for the increased distance in sandy soil (see Figs. 8.5 and 8.6), and however, the difference is very low as compared to silt loam soil.

Hence, it can be concluded that path loss is affected by frequency, depth, and distance, i.e., increase with increasing frequency, distance, and depth of antennas. This variation of frequency is due to soil permittivity. There are three major paths in underground–underground communication: (1) direct, (2) reflected, and (3) lateral. Direct wave is the line-of-sight path, reflected path is the one taken by wave when it is reflected from soil–air interface, and lateral path is taken because of the propagating along the soil–air interface. The reason for the path loss variation due to depth is because of multi-path effect. Direct path has no effect due to the depth increase, and however, it affects the lateral and reflected path. So when the depth is increased, path loss due to lateral and reflected waves increases, which in turn increases the overall path loss.

8.5.1.2 Impact of Soil Type on Attenuation

Figure 8.7 shows the comparison of attenuation in different soil types for varying depths and frequency. Figure 8.7a shows the result at the transmitter–receiver distance of 50 cm and Fig. 8.7b shows for the distance of 1 m. For 50 cm, sandy soil has 10–30 dB, 13 dB, 18 dB, and 20 dB lower path loss value at the depths of 10 cm, 20 cm, 30 cm, and 40 cm, respectively. Similarly, for 1 m, sandy soil has 12 dB, 20 dB, 18 dB, and 38 dB lower path loss value at the depths of 10 cm, 20 cm, 30 cm, and 40 cm, respectively.

It can be observed that, for both distances, path loss in sandy soil is less as compared to silt loam soil for all depths. The reason for this is because

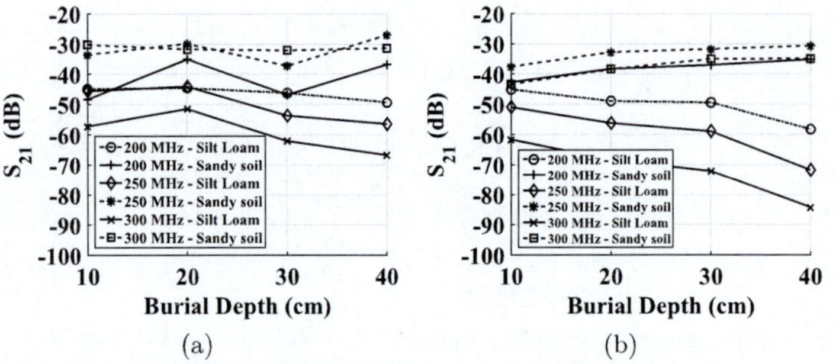

Fig. 8.7 (**a**) Comparison of attenuation in silt loam and sandy soils at **50** cm distance at different frequencies and (**b**) comparison of attenuation in silt loam and sandy soils at **1** distance at different frequencies

electromagnetic waves are absorbed by the bound water present in the soil and sandy soil absorbs less bound water as compared to silt loam soil. The water holding power of medium textured soil, e.g., silt loam and silty clay loam, is much higher than that of coarse soils such as sand and sandy loam. This is because coarse soils have a lower size of pores, hence low resistance against gravity and no aggregation.

8.5.1.3 Impact of Soil Moisture on Attenuation

This section evaluates the experiments done to study the effect of soil moisture on the signal attenuation. Different soil moisture levels, ranging from 0 to 50 CB, are used for these experiments. Figure 8.8 shows the experiment results at the transmitter–receiver distance of 50 cm in silt loam soil. It can be seen that, in Fig. 8.8a, there is a decrease of 5 dB as the moisture level goes from 0 to 50 and frequency increases. Figure 8.8b uses the frequency of 200 MHz to show the soil moisture effect at varying depth levels. There is a consistent path loss decrease at each depth value as the soil moisture level increases from 0 to 50 CB. For example, path loss decreases by 5 dB, 3 dB, 5 dB, and 3 dB at depths of 10 cm, 20 cm, 30 cm, and 40 cm, respectively, with increasing soil moisture. Figure 8.8c plots the path loss by keeping the fixed depth value, i.e., 10 cm, and varying frequency. For fixed depth of 10 cm, path loss decreases by 2 dB (at 250 MHz) and 3 dB (at 250 MHz).

Figure 8.9 shows the experiment results at the transmitter–receiver distance of 1m in silt loam soil. In Fig. 8.9a, a 3 dB decrease in path loss can be observed at 250 MHz as soil moisture goes from 0 to 50 CB. In Fig. 8.9c, experiments with varying frequency and a fixed depth of 10 cm show that path loss decreases by 2–3 dB as soil moisture goes from 0 to 50 CB. This difference increases further as frequency increases. Figure 8.9b uses the frequency of 200 MHz to show the soil moisture effect at varying depth levels. The trend is similar to the one observed in Fig. 8.8b, i.e., path loss decreases with increasing depth. For example, path loss

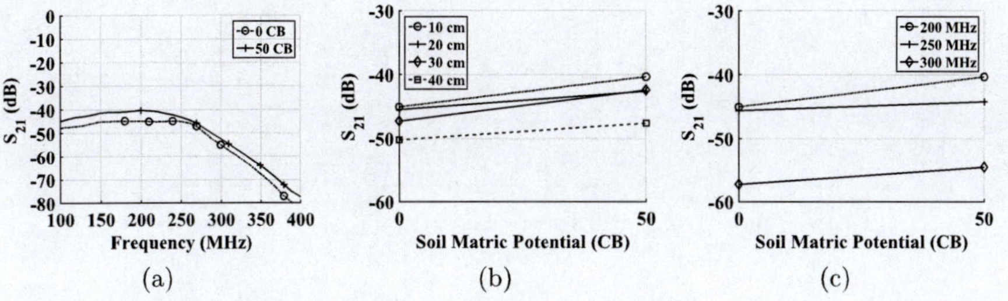

Fig. 8.8 Attenuation with soil moisture in silt loam soil at **50** cm distance: (**a**) S_{21}, (**b**) effect of change in soil moisture at different depths at 200 MHz frequency, and (**c**) effect of change in soil moisture at different frequencies at 10 cm depth

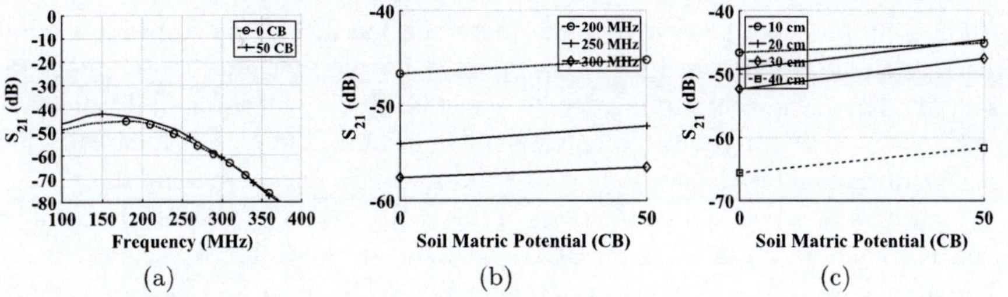

Fig. 8.9 Attenuation with soil moisture in silt loam soil at **1** distance: (**a**) S_{21}, (**b**) effect of change in soil moisture at different frequencies at 10 cm depth, and (**c**) effect of change in soil moisture at different depths at 200 MHz frequency

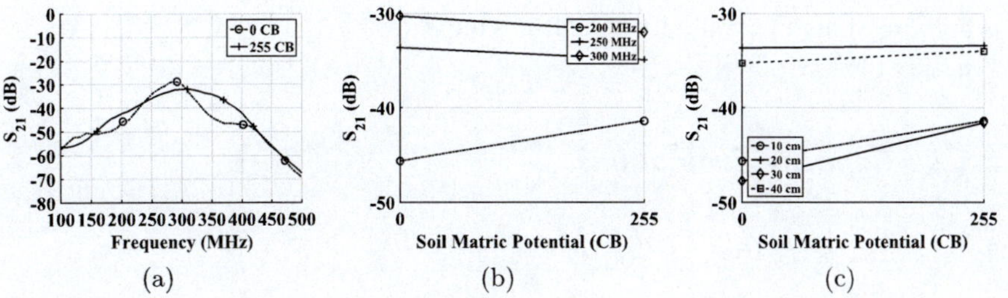

Fig. 8.10 Attenuation with soil moisture in sandy soil at **50** cm distance: (**a**) S_{21}, (**b**) effect of change in soil moisture at different frequencies at 10 cm depth, and (**c**) effect of change in soil moisture at different depths at 200 MHz frequency

decreases by 1 dB, 5 dB, 5 dB, and 4 dB at depths of 10 cm, 20 cm, 30 cm, and 40 cm, respectively, with increasing soil moisture.

Figures 8.10 and 8.11 show the effect of soil moisture on path loss for sandy soil at the transmitter–receiver distance of 50 cm and 1 m, respectively. Values of soil moisture considered for this experiment are in the range of 0 to 255 CB. In Fig. 8.10a, it can be seen that the path loss is minimum at 300 MHz, and after 300 MHz it starts increasing. Figure 8.10b shows that, for a fixed depth of

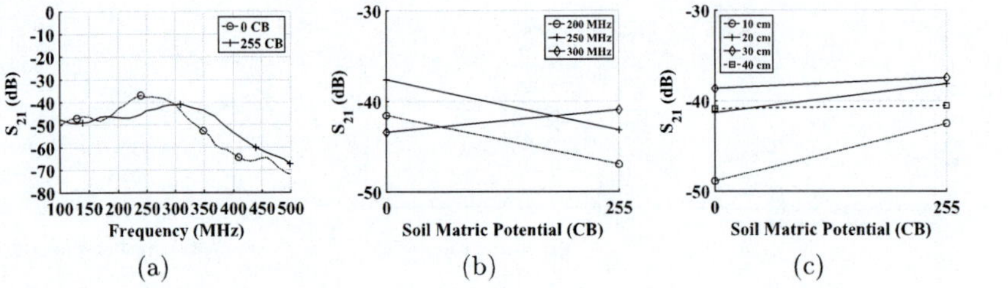

Fig. 8.11 Attenuation with soil moisture in sandy soil at **1** distance: (**a**) S_{21}, (**b**) effect of change in soil moisture at different frequencies at 10 cm depth, and (**c**) effect of change in soil moisture at different depths at 330 MHz frequency

10 cm, path loss shows negative trend (increasing) at 200 MHz and positive trend (decreasing) at 250 MHz and 300 MHz. Figure 8.10c plots the effect of soil moisture with fixed frequency but varying depths. It can be observed that for all depth levels path loss is increasing with decreasing soil moisture. The reason for this is the change of resonant frequency to high spectrum with decreasing soil moisture.

Figure 8.11a shows similar trends as of Fig. 8.10a, and however, for 1 m distance path is minimum at 250 MHz at 0 CB soil moisture level and it is minimum at 300 MHz at 255 CB soil moisture level. In Fig. 8.11b, path loss is decreasing at 200 and 250 MHz and decreasing at 300 MHz as soil moisture goes from 0 to 250 CB. Figure 8.11c shows the decrease in path loss for all depths at 200 MHz.

These experiments conclude that soil moisture is inversely proportional to path loss, i.e., path loss increases with decrease in soil moisture for both soil types. This is because of high permittivity at high soil moisture.

8.5.2 Capacity Evaluations

8.5.3 Power Delay Profile Measurements

Figure 8.12 shows the result of power delay profiling (PDP) for the experiments at 50 cm and 1 m.

The resulting power delay profiles are shown in Fig. 8.12. The speed of the wave in soil is given as: $S = c/n$, where n is the refractive index and c is the speed of light 3×10^8 m/s. Since the permittivity of soil is a complex number, the refractive index of soil is calculated as follows:

$$n = \sqrt{\frac{\sqrt{\epsilon'^2 + \epsilon''^2} + \epsilon'}{2}}, \tag{8.2}$$

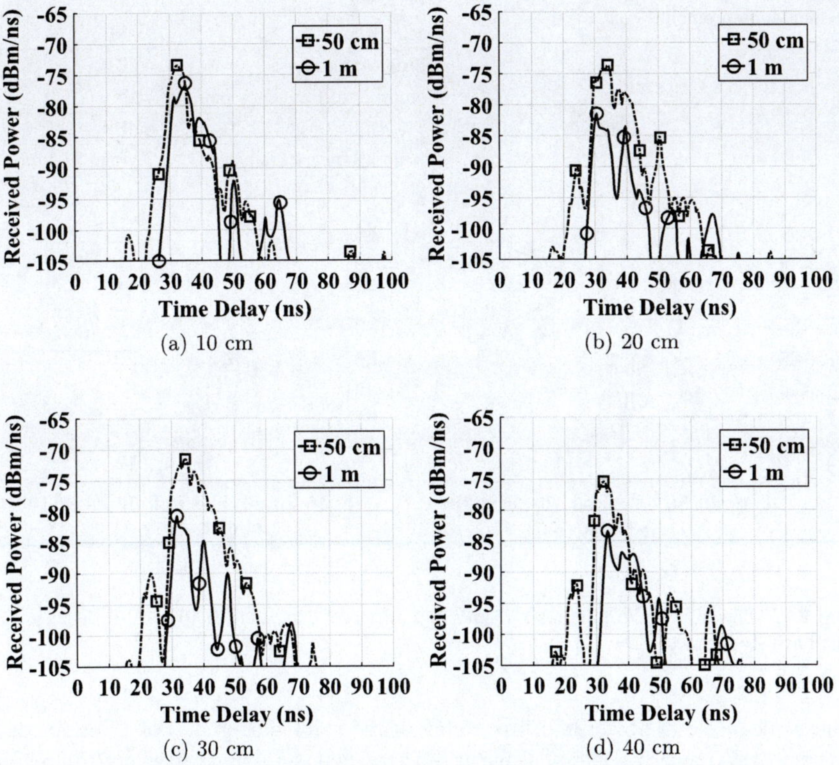

Fig. 8.12 Power Delay Profiles (PDPs) measured at 50 cm and 1 m distances, at different depths in silt loam soil at near-saturation: **(a)** 10 cm, **(b)** 20 cm, **(c)** 30 cm, and **(d)** 40 cm

where ϵ' and ϵ'' are the real and imaginary parts of the relative permittivity of the soil.

The wave speed is calculated in silt loam soil using refractive index. Refractive index is calculated as $5.6 \times 10_7$ on the basis of properties given in Table 8.3. It is five times slower and 19% of the speed of light.

Figure 8.12 shows the results from the experiments done at the distance of 50 cm and 1 m for all depths. The first multipath component is the direct wave. It is present at the delay of 18–28 ns at 50 cm and does not appear for 1 m distance. The reason for this is that direct wave is more attenuated at 1 m than 50 cm. Lateral waves can be seen at the time delay of 30–40 ns as the strongest component among all PDPs. The delay for lateral wave is similar for both, 1 m and 50 cm, because it propagates faster in air. In general, the power of lateral wave is 10–15 dB faster than direct wave.

Figure 8.13 compares PDP for all four depths at the distances of 50 cm (see Fig. 8.13a) and 1 m (see Fig. 8.13b) between transmitter and receiver. It can be observed from the results that if the distance is kept the same, increase in depth

Table 8.3 Underground channel measurement parameters

Parameter	Value
Start frequency	10 MHz
Stop frequency	4 GHz
Number of frequency points	401
Transmit power	5 dBm
Vector network analyzer	Agilent FieldFox

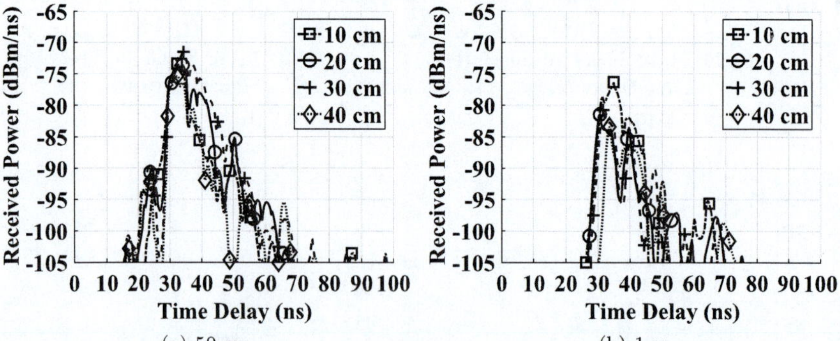

Fig. 8.13 Power Delay Profile in silt loam soil at different depths at (**a**) 50 cm T–R distance and (**b**) 1 m T–R distance

causes decrease in power and time delay of the received signal. For example, at 1 m, the peak power of lateral wave is −75 dB and −83 dB at the depth of 10 cm and 40 cm, respectively. Similarly, the same trend is seen for the time delay, i.e., in Fig. 8.13a lateral wave arrives at 29 ns and 32 ns at the depths of 10 cm and 40 cm, respectively.

Figure 8.14 shows the PDP measurement results at distances of 50 cm and 1 m and a depth of 20 cm. The effect of soil moisture is measured by varying the soil moisture values for varying soil moisture values. For both distances, the received signal strength is increasing with decreasing soil moisture. An important observation to note is that direct component vanishes with increasing distance due to higher soil attenuation.

Figure 8.15 shows the measurement of PDP with different types of soil. It is observed that, due to its low water holding capacity, the power of received signal is the highest in sandy soil as compared to silt loam and clay loam soil.

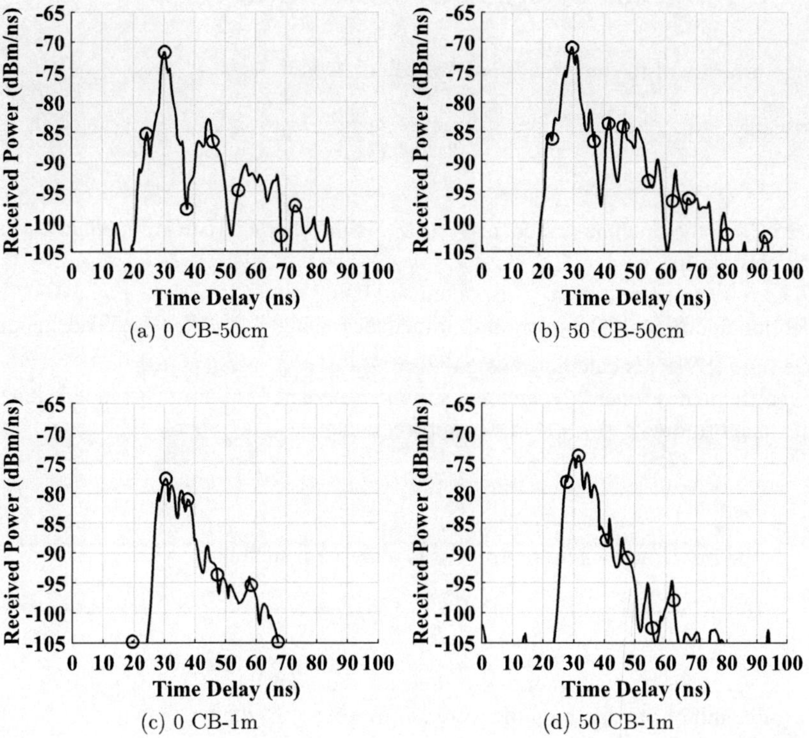

Fig. 8.14 Power Delay Profiles (PDPs) measured at 50 cm and 1 m distances, at 20 cm depths for different soil moisture levels: (**a**) 0 CB-50 cm, (**b**) 50 CB-50 cm, (**c**) 0 CB-1 m, and (**d**) 50 CB-1 m

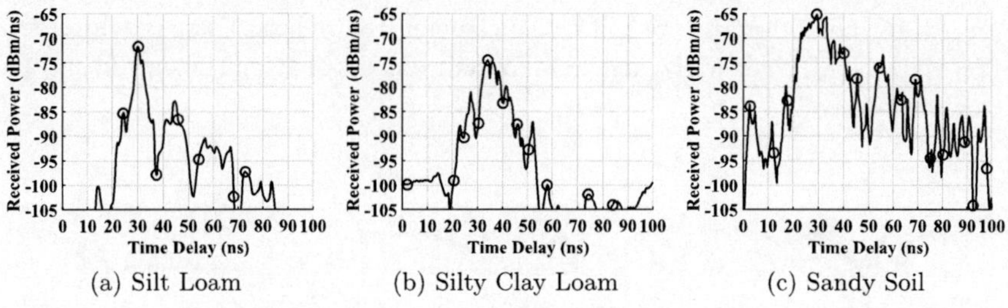

Fig. 8.15 Power Delay Profiles (PDPs) measured in different soils: (**a**) silt loam, (**b**) silty clay loam, and (**c**) sandy soil

8.6 UG Antenna Return Loss Measurements

The return loss of the antenna (in dB) is calculated as

$$RL_{dB} = 20\log_{10}\left|\frac{Z_a - Z_0}{Z_a + Z_0}\right|, \tag{8.3}$$

where Z_a is the antenna impedance and Z_0 is the characteristics impedance of the transmission line.

The reflection coefficient Γ is calculated using return loss as $|\Gamma| = 10^{\frac{RL}{20}}$. The reflection coefficient is converted to impedance using $Z_a = Z_0\frac{1+\Gamma}{1-\Gamma}$. The standing wave ratio (SWR) is calculated as $SWR = \frac{1+|\Gamma|}{1-|\Gamma|}$.

The frequency where the antenna's input impedance is pure resistance is known as resonant frequency (f_r). At resonant frequency f_r,

$$Z_a|_{f=f_r} = Z_r = R_a. \tag{8.4}$$

It is the frequency where return loss is maximum such that

$$f_r = \max(RL_{dB}). \tag{8.5}$$

The return loss experiments are done to measure the effect of soil type, soil moisture, and burial depth on the resonant frequency of the UG antenna.

8.6.1 *Effects of Soil Type*

Figure 8.16 shows the effect of soil type on antenna return loss. There is a difference of 63 MHz, 59 MHz, 55 MHz, and 43 MHz at the depths of 10 cm, 20 cm, 30 cm, and 40 cm, respectively.

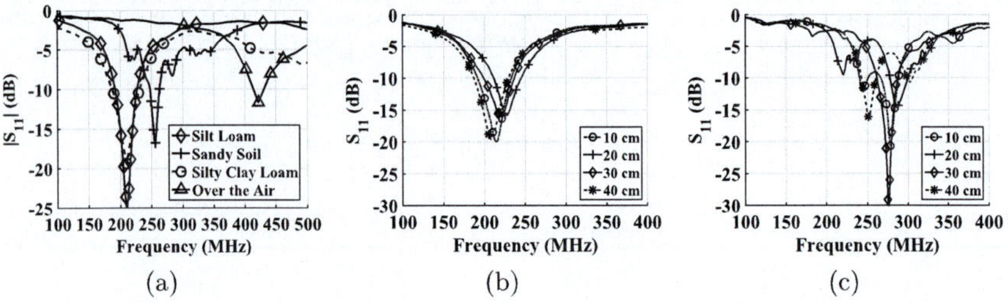

Fig. 8.16 Change in return loss in different soil: (**a**) return loss at all four depths in sandy and silt loam soil: (**b**) silt loam, and (**c**) sandy soil

8.6.2 *Impact of Change in Soil Moisture*

Figure 8.17 shows different parameters affecting the return loss of antenna. It uses the soil matric potential values from 0 to 50 CB for all experiments. The soil used for this experiment is silt loam soil. Figure 8.17a shows the return loss for 10 cm depth at varying frequencies. It can be seen in Fig. 8.17b, for soil moisture values of 0 CB and 50 CB, that resonant frequency changes from 211 MHz to 219 MHz at the depths of 10 cm, 221 MHz to 227 MHz at the depth of 20 cm, 221 MHz to 231 MHz at the depth of 30 cm, and 201 MHz to 231 MHz at the depth of 40 cm.

In Fig. 8.17c, at shifted frequency, the reflection coefficient changes from −15 dB to −17 dB at the depth of 10 cm, −18 dB to −15 dB at the depth of 20 cm, −20 dB to −17 dB at the depth of 30 cm, and −20 dB to −18 dB at the depth of 40 cm. Figure 8.17c plots the change in antenna bandwidth at soil moisture values of 0 CB and 255 CB. It can be observed that with decreasing soil moisture bandwidth changes from 40 MHz to 29 MHz at the depth of 10 cm, 30 MHz to 34 MHz at the

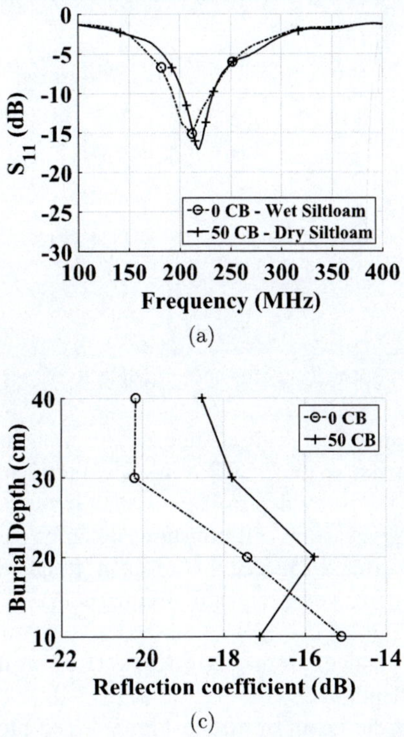

Fig. 8.17 Return loss in silt loam soil: (**a**) S_{11} at different frequencies, (**b**) change in resonant frequency with burial depth, (**c**) reflection coefficient (dB) at different burial depths, and (**d**) antenna bandwidth at different burial depths

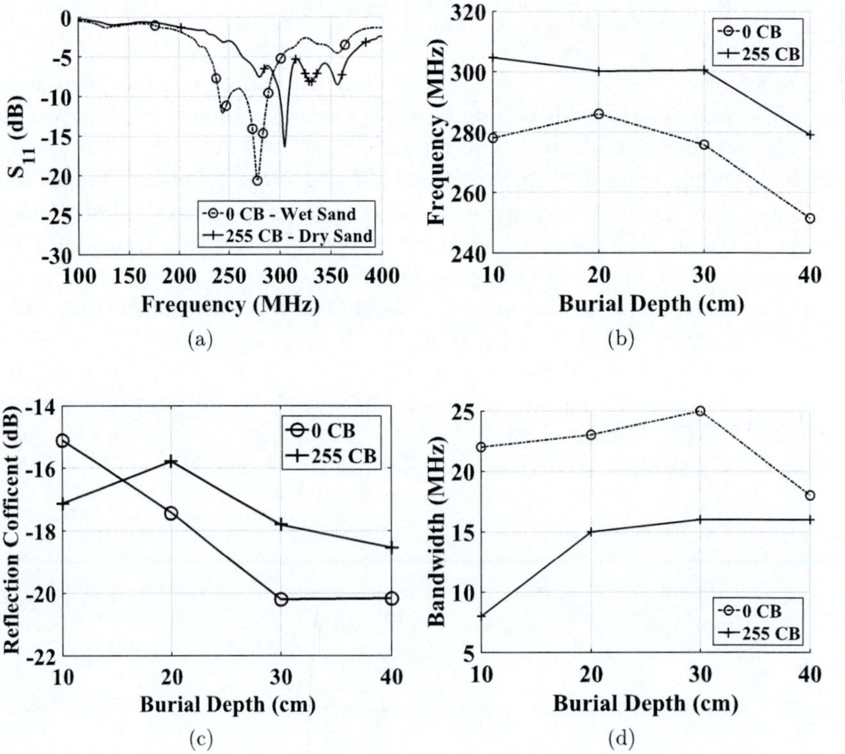

Fig. 8.18 Return loss in sandy soil: (**a**) S_{11} at different frequencies, (**b**) change in resonant frequency with burial depth, (**c**) reflection coefficient (dB) at different burial depths, (**d**) antenna bandwidth at different burial depths

depth of 20 cm, 39 MHz to 34 MHz at the depth of 30 cm, and 37 MHz to 30 MHz at the depth of 40 cm.

Figure 8.18 plots the return loss for sandy soil with change in soil moisture. It uses the soil matric potential values from 0 to 255 CB for all experiments. Figure 8.18a shows the antenna return loss for 10 cm depth. The resonant frequency changes from 278 MHz to 305 MHz with decrease in soil moisture. It can be seen in Fig. 8.18b, for soil moisture values of 0 CB and 255 CB, that resonant frequency changes from 276 MHz to 301 MHz at the depths of 20 cm and 30 cm and 251 MHz to 279 MHz at the depth of 40 cm.

In Fig. 8.18c, the reflection coefficient changes from −20 dB to −16 dB at the depth of 10 cm, −14 dB to −12 dB at the depth of 20 cm, −31 dB to −15 dB at the depth of 30 cm, and −16 dB to −15 dB at the depth of 40 cm. Figure 8.18d plots the change in antenna bandwidth at soil moisture values of 0 CB and 255 CB. It can be observed that as the soil moisture decreases bandwidth changes from 22 MHz to 8 MHz at the depth of 10 cm, 23 MHz to 15 MHz at the depth of 20 cm, 25 MHz to 16 MHz at the depth of 30 cm, and 18 MHz to 16 MHz at the depth of 40 cm.

The analysis has shown that antenna return loss is affected by soil moisture for sandy and silt loam soils. As the soil moisture increases, the resonant frequency goes to a lower range. Contrary to over-the-air communications, resonant frequency and optimal frequency (where the maximum capacity is achieved) of antenna are different. Finding an optimal frequency for UG antenna in soil is a potential area to work on.

8.6.3 Impact of Burial Depth

Figure 8.16 shows the effect of burial depth on return loss in sandy soil (Fig. 8.16c) and silt loam (Fig. 8.16b). The chosen depths are 10 cm, 20 cm, 30 cm, and 40 cm. Resonant frequency increases from 215 MHz (at 10 cm depth) to 227 MHz (at 20 cm depth) and decreases to 220 MHz (at 30 cm) and 208 MHz (at 40 cm). Reflection coefficient is −19.92 dB, −15.76 dB, −16.04 dB, and −19.57 dB at the depths of 10 cm, 20 cm, 30 cm, and 40 cm, respectively. To analyze the bandwidth of antenna, −10 dB is chosen as the threshold power value. The bandwidth is measured as 40 MHz, 32 MHz, 37 MHz, and 42 MHz at the depths of 10 cm, 20 cm, 30 cm, and 40 cm, respectively.

Figure 8.16c shows the effect of sandy soil on return loss. As in the case of silt loam soil, resonant frequency increases form 278 MHz (at 10 cm depth) to 286 MHz (at 20 cm depth) and then decreases to 275 MHz (at 30 cm) and 251 MHz (at 40 cm). Reflection coefficient is −19.92 dB, −15.76 dB, −16.04 dB, and −19.57 dB at the depths of 10 cm, 20 cm, 30 cm, and 40 cm, respectively. Reflection coefficient is −20.66 dB (for 10 cm depth), −14.57 dB (for 20 cm depth), −32 dB (for 30 cm depth), and −16.07 dB (for 40 cm depth). The bandwidth was measured as 22 MHz, 21 MHz, 26 MHz, and 17 MHz at the depths of 10 cm, 20 cm, 30 cm, and 40 cm, respectively.

It is very important to characterize underground channel models to design IOUT applications [1, 4, 6–26, 42–51, 53–55]. In this chapter, the design of different types of testbeds for underground channel modeling is presented. It also presented a detailed measurement campaign which was completed over the several years. Empirical results for the underground antenna performance and underground channel transfer functions were presented using different types of soil (sandy and silt loam). Experiments show that indoor testbed gives speedy, efficient, and improved characterization of underground communications. From experiments, it is shown how soil moisture, soil texture, and burial depth of the antenna affect the performance of underground channel and communication. These empirical measurements can act as the preliminary step in designing the next generate IOUT communication protocols for the development of improved wireless underground communications.

References

1. A. Konda, A. Rau, M.A. Stoller, J.M. Taylor, A. Salam, G.A. Pribil, C. Argyropoulos, S.A. Morin, Soft microreactors for the deposition of conductive metallic traces on planar, embossed, and curved surfaces. Adv. Funct. Mat. **28**(40), 1803020 (2018)
2. U. Raza, A. Salam, On-site and external power transfer and energy harvesting in underground wireless. Electronics **9**(4) (2020)
3. U. Raza, A. Salam, A survey on subsurface signal propagation. Smart Cities **3**(4), 1513–1561 (2020)
4. U. Raza, A. Salam, Wireless underground communications in sewer and stormwater overflow monitoring: radio waves through soil and asphalt medium. Information **11**(2) (2020)
5. U. Raza, A. Salam, Zenneck waves in decision agriculture: an empirical verification and application in EM-based underground wireless power transfer. Smart Cities **3**(2), 308–340 (2020)
6. A. Salam, Pulses in the sand: Long range and high data rate communication techniques for next generation wireless underground networks. ETD collection for University of Nebraska - Lincoln (AAI10826112) (2018)
7. A. Salam, A comparison of path loss variations in soil using planar and dipole antennas, in *2019 IEEE International Symposium on Antennas and Propagation* (IEEE, Piscataway, 2019)
8. A. Salam, Design of subsurface phased array antennas for digital agriculture applications, in *Proceedings of the 2019 IEEE International Symposium on Phased Array Systems and Technology (IEEE Array 2019)*, Waltham (2019)
9. A. Salam, A path loss model for through the soil wireless communications in digital agriculture, in *2019 IEEE International Symposium on Antennas and Propagation* (IEEE, Piscataway, 2019)
10. A. Salam, Sensor-free underground soil sensing, in *ASA, CSSA and SSSA International Annual Meetings (2019)*. ASA-CSSA-SSSA (2019)
11. A. Salam, Subsurface MIMO: a beamforming design in internet of underground things for digital agriculture applications. J. Sens. Actuator Netw. **8**(3) (2019)
12. A. Salam, *Underground Environment Aware MIMO Design Using Transmit and Receive Beamforming in Internet of Underground Things* (Springer International Publishing, Cham, 2019), pp. 1–15
13. A. Salam, An underground radio wave propagation prediction model for digital agriculture. Information **10**(4) (2019)
14. A. Salam, Underground soil sensing using subsurface radio wave propagation, in *5th Global Workshop on Proximal Soil Sensing*, COLUMBIA (2019)
15. A. Salam, *Internet of Things for Environmental Sustainability and Climate Change* (Springer International Publishing, Cham, 2020), pp. 33–69
16. A. Salam, *Internet of Things for Sustainability: Perspectives in Privacy, Cybersecurity, and Future Trends* (Springer International Publishing, Cham, 2020), pp. 299–327
17. A. Salam, *Internet of Things for Sustainable Community Development*, 1st edn. (Springer Nature, Berlin, 2020)
18. A. Salam, *Internet of Things for Sustainable Community Development: Introduction and Overview* (Springer International Publishing, Cham, 2020), pp. 1–31
19. A. Salam, *Internet of Things for Sustainable Forestry* (Springer International Publishing, Cham, 2020), pp. 147–181
20. A. Salam, *Internet of Things for Sustainable Human Health* (Springer International Publishing, Cham, 2020), pp. 217–242
21. A. Salam, *Internet of Things for Sustainable Mining* (Springer International Publishing, Cham, 2020), pp. 243–271
22. A. Salam, *Internet of Things for Water Sustainability* (Springer International Publishing, Cham, 2020), pp. 113–145

23. A. Salam, *Internet of Things in Agricultural Innovation and Security* (Springer International Publishing, Cham, 2020), pp. 71–112
24. A. Salam, *Internet of Things in Sustainable Energy Systems* (Springer International Publishing, Cham, 2020), pp. 183–216
25. A. Salam, *Internet of Things in Water Management and Treatment* (Springer International Publishing, Cham, 2020), pp. 273–298
26. A. Salam, U. Karabiyik, A cooperative overlay approach at the physical layer of cognitive radio for digital agriculture, in *Third International Balkan Conference on Communications and Networking 2019 (BalkanCom'19)*, Skopje (2019)
27. A. Salam, U. Raza, *Autonomous Irrigation Management in Decision Agriculture* (Springer International Publishing, Cham, 2020), pp. 379–398
28. A. Salam, U. Raza, *Current Advances in Internet of Underground Things* (Springer International Publishing, Cham, 2020), pp. 321–356
29. A. Salam, U. Raza, *Decision Agriculture* (Springer International Publishing, Cham, 2020), pp. 357–378
30. A. Salam, U. Raza, *Electromagnetic Characteristics of the Soil* (Springer International Publishing, Cham, 2020), pp. 39–59
31. A. Salam, U. Raza, *Modulation Schemes and Connectivity in Wireless Underground Channel* (Springer International Publishing, Cham, 2020), pp. 125–166
32. A. Salam, U. Raza, On burial depth of underground antenna in soil horizons for decision agriculture, in *2020 International Conference on Internet of Things (ICIOT-2020)*, Honolulu (2020)
33. A. Salam, U. Raza, *Signals in the Soil*, 1 edn. (Springer Nature, Berlin, 2020)
34. A. Salam, U. Raza, *Signals in the Soil: An Introduction to Wireless Underground Communications* (Springer International Publishing, Cham, 2020), pp. 3–38
35. A. Salam, U. Raza, *Signals in the Soil: Subsurface Sensing* (Springer International Publishing, Cham, 2020), pp. 251–297
36. A. Salam, U. Raza, *Signals in the Soil: Underground Antennas* (Springer International Publishing, Cham, 2020), pp. 189–215
37. A. Salam, U. Raza, *Soil Moisture and Permittivity Estimation* (Springer International Publishing, Cham, 2020), pp. 299–317
38. A. Salam, U. Raza, *Underground Phased Arrays and Beamforming Applications* (Springer International Publishing, Cham, 2020), pp. 217–248
39. A. Salam, U. Raza, *Underground Wireless Channel Bandwidth and Capacity* (Springer International Publishing, Cham, 2020), pp. 167–188
40. A. Salam, U. Raza, *Variable Rate Applications in Decision Agriculture* (Springer International Publishing, Cham, 2020), pp. 399–423
41. A. Salam, U. Raza, *Wireless Underground Channel Modeling* (Springer International Publishing, Cham, 2020), pp. 61–121
42. A. Salam, S. Shah, Internet of things in smart agriculture: Enabling technologies, in *2019 IEEE 5th World Forum on Internet of Things (WF-IoT) (WF-IoT 2019)*, Limerick (2019)
43. A. Salam, M.C. Vuran, Impacts of soil type and moisture on the capacity of multi-carrier modulation in internet of underground things, in *Proceedings of the 25th ICCCN 2016*, Waikoloa (2016)
44. A. Salam, M.C. Vuran, Smart underground antenna arrays: a soil moisture adaptive beamforming approach, in *Proceedings of the IEEE INFOCOM 2017*, Atlanta (2017)
45. A. Salam, M.C. Vuran, Wireless underground channel diversity reception with multiple antennas for internet of underground things, in *Proceedings of the IEEE ICC 2017*, Paris (2017)
46. A. Salam, M.C. Vuran, EM-Based Wireless Underground Sensor Networks, in *Underground Sensing*, ed. by S. Pamukcu, L. Cheng (Academic Press, Cambridge, 2018), pp. 247–285
47. A. Salam, M.C. Vuran, S. Irmak, Pulses in the sand: Impulse response analysis of wireless underground channel, in *The 35th Annual IEEE International Conference on Computer Communications (INFOCOM 2016)*, San Francisco (2016)

48. A. Salam, M.C. Vuran, S. Irmak, Towards internet of underground things in smart lighting: a statistical model of wireless underground channel, in *Proceedings of the 14th IEEE International Conference on Networking, Sensing and Control (IEEE ICNSC)*, Calabria (2017)
49. A. Salam, M.C. Vuran, S. Irmak, Di-sense: in situ real-time permittivity estimation and soil moisture sensing using wireless underground communications. Comput. Netw. **151**, 31–41 (2019)
50. A. Salam, M.C. Vuran, X. Dong, C. Argyropoulos, S. Irmak, A theoretical model of underground dipole antennas for communications in internet of underground things. IEEE Trans. Antennas Propag. **67**(6), 3996–4009 (2019)
51. A. Salam, A.D. Hoang, A. Meghna, D.R. Martin, G. Guzman, Y.H. Yoon, J. Carlson, J. Kramer, K. Yansi, M. Kelly et al., The future of emerging IoT paradigms: architectures and technologies (2019, preprint)
52. A. Salam, M.C. Vuran, S. Irmak, A statistical impulse response model based on empirical characterization of wireless underground channel. IEEE Trans. Wireless Commun. **19** (2020)
53. S. Temel, M.C. Vuran, M.M.R. Lunar, Z. Zhao, A. Salam, R.K. Faller, C. Stolle, Vehicle-to-barrier communication during real-world vehicle crash tests. Comput. Commun. **127**, 172–186 (2018)
54. M.C. Vuran, A. Salam, R. Wong, S. Irmak, Internet of underground things in precision agriculture: architecture and technology aspects. Ad Hoc Netw. **81**, 160–173 (2018)
55. M.C. Vuran, A. Salam, R. Wong, S. Irmak, Internet of underground things: sensing and communications on the field for precision agriculture, in *2018 IEEE 4th World Forum on Internet of Things (WF-IoT) (WF-IoT 2018)*, Singapore (2018)

Index

A. Salam, *Internet of Things in Smart Sewer and Drainage Systems*,
https://doi.org/10.1007/978-3-031-48109-3